Reinforced Concrete Masonry Construction

Inspector's Handbook

SECOND EDITION

Reinforced Concrete Masonry Construction
Inspector's Handbook

SECOND EDITION

James E. Amrhein, S.E.

Michael W. Merrigan

Published by

Masonry Institute of America

2550 Beverly Boulevard
Los Angeles, California 90057

in cooperation with

International Conference of Building Officials

5360 South Workman Mill Road
Whittier, California 90601

2nd Edition

Copyright 1987, 1989 by
Masonry Institute of America
and
International Conference of Building Officials

Library of Congress Catalog Card Number 87-061825
ISBN 0-940116-09-X

Printed in the United States of America

Portions of this publication are reproduced, with permission, from the Uniform Building Code, 1988 edition and the Uniform Building Code Standards, 1988 edition copyright © International Conferance of Building Officials.

This handbook was prepared in keeping with current information and practice for the present state of the art of masonry design and construction.

MIA 206-10-87-5M
 3-89-5M

Table of Contents

Acknowledgements xv

Masonry Institute of America xvi

International Conference of Building Officials xviii

SECTION 1. INTRODUCTION 1

1.1 **General** 1

1.2 **The Inspector** 2

1.3 **Responsibilities and Duties** 2

1.4 **Equipment and Materials for the Inspector** 3

1.5 **Terminology** 4

SECTION 2. MATERIALS 11

2.1 **General** 11

2.2 **Material Standards** 11

2.3   **Concrete Masonry Units**   14

    2.3.1   General   14

    2.3.2   Dimensions   15

    2.3.3   Wide Selection of Units   15

    2.3.4   Component Units   16

    2.3.5   Storing Masonry Units   16

2.4   **Cementitious Materials**   19

    2.4.1   Portland Cement   19

    2.4.2   Plastic Cement   20

    2.4.3   Mortar Cement   20

    2.4.4   Masonry Cement   21

    2.4.5   Lime   21

2.5   **Aggregates for Mortar and Grout**   23

2.6   **Reinforcing Steel**   23

    2.6.1   General   23

    2.6.2   Reinforcing Bars   23

    2.6.3   Identification Marks   26

    2.6.4   Overall Bar Diameters   27

2.7   **Joint Reinforcing Steel**   27

    2.7.1   General   27

    2.7.2   Description   28

    2.7.3   Configuration and Size of
Longitudinal and Cross Wires   29

    2.7.4   Material Requirements   29

    2.7.5   Fabrication   30

2.8   **Water**   30

2.9   Additives and Admixtures   30

2.10 Mortar   31

     2.10.1   General   31

     2.10.2   Proportions of Mortar   32

     2.10.3   Mortar Aggregate—Sand   35

     2.10.4   Mixing   36

     2.10.5   Retempering   37

     2.10.6   Color   38

     2.10.7   Proprietary Mortars   39

     2.10.8   Mortar Admixtures   39

2.11 Grout   39

     2.11.1   General   39

     2.11.2   Types of Grout   40

     2.11.3   Proportions   40

     2.11.4   Aggregate for Grout   42

     2.11.5   Mixing   42

     2.11.6   Grout Admixtures   43

     2.11.7   Anti-Freeze Compounds   44

**SECTION 3. QUALITY CONTROL,
               SAMPLING AND TESTING**   45

3.1   Quality Control   45

3.2   Sampling and Testing   46

     3.2.1   Cone Penetration Test for
           Consistency of Mortar   46

     3.2.2   Field Test for Mortar Strength   48

	3.2.3	Field Compressive Test Specimens for Mortar	48
	3.2.4	Mortar Strength Requirements	50
	3.2.5	Field Test for Grout	52
	3.2.6	Field Compressive Test Specimens for Grout	52
	3.2.7	Grout Strength Requirements	55
3.3	**Concrete Masonry Units**		56
3.4	**Prism Testing**		57
	3.4.1	General	57
	3.4.2	Standard Prism Test	59
	3.4.3	Tests of Masonry Prisms	59
	3.4.4	Specified Compressive Strength, f'_m	61
3.5	**Summary**		63

SECTION 4. CONSTRUCTION ... 77

4.1	**General**		77
4.2	**Preparation of Foundation and Site**		77
4.3	**Materials, Handling, Storage and Preparation**		78
4.4	**Placement and Layout**		83
	4.4.1	General	83
	4.4.2	Placing Masonry Units	85
	4.4.3	Typical Layout of CMU Walls	88
		A. Corner Details	88
		B. Pilaster Details	92

C. Connections of Intersecting
    Walls and Embedded Columns 95

D. Lintel and Bond Beam 96

E. Arrangement of Open End Units 97

F. Wall Assembly 100

4.5 Mortar Joints 102

4.6 Reinforcing Steel 107

4.6.1 Maximum Size and Amount of
       Reinforcing Steel 107

4.6.2 Spacing of Steel in Walls 107

4.6.3 Clearances of Steel and Masonry 110

4.6.4 Securing Reinforcing Steel 114

4.6.5 Location Tolerances of Bars 116

4.6.6 Lap Splices, Reinforcing Bars 116

4.6.7 Lap Splices, Joint Reinforcing 119

4.6.8 Coverage and Layout of
       Joint Reinforcing Steel 120

       A. Coverage 120

       B. Layout 123

4.6.9 Hooks and Bends in Reinforcing Bars 125

4.6.10 Anchorage of Shear Reinforcing Steel 127

4.7 Column Reinforcement 132

4.7.1 Vertical Reinforcement 132

4.7.2 Reinforcing Tie Details 134

       A. Lateral Tie Details 134

       B. Lateral Tie Spacing—
           Seismic Zones 0, 1 and 2 138

C. Lateral Tie Spacing—
Seismic Zones 3 and 4 139

D. Layout of Ties in
Concrete Masonry Columns 140

4.7.3 Projection Wall Columns or Pilasters 142

4.7.4 Flush Wall Columns or Pilasters 144

4.7.5 Ties on Compression Steel in Beams 145

4.7.6 Anchor Bolts 145

A. Anchor Bolt Clearance 145

B. Anchor Bolt Ties 146

C. Anchor Bolts in Walls 146

D. Embedment of Anchor Bolts 149

**4.8 Special Provisions for Seismic
Design and Construction** 153

4.8.1 General 153

4.8.2 Seismic Zones Nos. 0 and 1 154

4.8.3 Seismic Zone No. 2 156

4.8.4 Seismic Zones Nos. 3 and 4 157

4.8.5 Shear Walls 160

4.9 Grouting of Masonry Walls 162

4.9.1 General 162

4.9.2 Mortar Protrusions 163

4.9.3 Grout Slump 164

4.9.4 Grouting Limitations 164

4.9.5 Low-Lift Grouting 166

4.9.6 Cleanouts 168

4.9.7   High-Lift Grouting 168

　　　A.　U.B.C. Requirements 168

　　　B.　O.S.A. Requirements 172

4.9.8   Consolidation of Grout 180

4.9.9   Fluid Mortar for Grout 181

4.9.10 Grout Barriers 182

4.9.11 Use of Aluminum Equipment 183

4.9.12 Pumping Grout 183

4.10   Bracing of Walls 184

4.11   Pipes and Conduits Embedded in Masonry 185

4.12   Adjacent Work 185

4.13   Intersecting Structural Elements 186

4.13.1 Wall to Wall 186

4.13.2 Walls to Floor or Roof 186

4.14   Multi-Wythe Walls 190

4.14.1 General 190

4.14.2 Metal Ties for Cavity Wall Construction 191

4.14.3 Metal Ties for Grouted
　　　 Multi-Wythe Construction 192

4.14.4 Joint Reinforcing 193

4.14.5 Stack Bond 193

4.15   Crack Control 193

4.15.1 Jointing; Control Joints
　　　 and Expansion Joints 195

4.15.2 Control Joints 196

4.15.3 Expansion joints 197

4.15.4 Summary 199

4.15.5 Crack Repair 199

4.16 Cold Weather Masonry Construction 200

4.16.1 General 200

4.16.2 Preparation 200

4.16.3 Construction. 201

4.16.4 Protection. 201

4.16.5 Placing Grout and
Protection of Grouted Masonry 202

4.16.6 Summary of Recommended
Cold Weather Practices 202

4.17 Hot Weather Masonry Construction 203

4.17.1 General 203

4.17.2 Performance 204

4.17.3 Handling and Selection of Materials 204

4.17.4 Construction Procedure. 205

4.18 Wet Weather Masonry Construction 205

4.18.1 General 205

4.18.2 Performance 205

4.18.3 Construction Procedures 206

4.18.4 Protection of Masonry 206

**4.19 Reinforced Concrete Masonry
Inspection Checklist** 207

SECTION 5. MASONRY UNITS 217

**5.1 I.C.B.O. Evaluation Service, Inc.
Evaluation Reports** 217

5.2 **Typical Concrete Masonry Units** 217

    5.2.1   Precision Units 217

    5.2.2   Slumped Block 217

    5.2.3   Custom Face Units 234

    5.2.4   Split Face Units 234

    5.2.5   Special Proprietary Units 234

5.3 **Length, Height and Quantities**
in Concrete Masonry Walls 237

    5.3.1   Length and Height 237

    5.3.2   Quantities of Materials 240

SECTION 6. GLOSSARY OF TERMS 243

SECTION 7. UNIFORM BUILDING
                   CODE STANDARDS 267

Standard No. 24–3 Concrete Building Brick 267

Standard No. 24–4 Hollow and Solid Load-Bearing
Concrete Masonry Units 271

Standard No. 24–6 Non-Load-Bearing
Concrete Masonry Units 277

Standard No. 24–7 Sampling and Testing
Concrete Masonry Units 281

Standard No. 24–15 Joint Reinforcement for Masonry . . . 286

Standard No. 24–16 Cement, Masonry 293

Standard No. 24–17 Quicklime for Structural Purposes . . . 294

Standard No. 24–18 Hydrated Lime for
Masonry Purposes 295

Standard No. 24–20 Mortar for Unit Masonry and
Reinforced Masonry other than Gypsum 298

Standard No. 24–21 Aggregate for Masonry Mortar 303

Standard No. 24–22 Field Test Specimens for Mortar 305

Standard No. 24–23 Aggregate for Masonry Grout 306

Standard No. 24–26 Test Method for Compressive
Strength of Masonry Prisms 308

Standard No. 24–27 Standard Test Method for Drying
Shrinkage of Concrete Block 313

Standard No. 24–28 Method of Sampling
and Testing Grout 323

Standard No. 24–29 Grout for Masonry 326

SECTION 8. REFERENCES 329

Acknowledgements

The Masonry Institute of America appreciates the review and suggestions received by the California Council of Construction Inspectors Association and its Board of Registered Construction Inspectors and by Stuart R. Beavers, Executive Director of the Concrete Masonry Association of California and Nevada. The review and constructive suggestions of Donald A. Wakefield, masonry consultant and past president of The Masonry Society, is recognized and appreciated. Their comments and advice have helped improve this handbook.

The Masonry Institute of America appreciates the use of the 4th edition of the CMACN "Reinforced Concrete Masonry Inspector's Manual" which served as a source document.

James E. Amrhein, S.E.
Executive Director
Masonry Institute of America
Los Angeles, California

Masonry Institute of America

The Masonry Institute of America, founded in 1957 under the name Masonry Research, is a promotion, technical research organization established to improve and extend the use of masonry. Supported by the mason contractors through a labor management contract between the unions and contractors, Masonry Institute of America is active in Los Angeles County in promoting new ideas and masonry work, improving building codes, conducting research projects, presenting design, construction and inspection seminars, and writing technical and non-technical papers, all for the purpose of improving the masonry industry.

Masonry Institute of America does not engage in the practice of architectural or engineering design or construction nor does it sell masonry materials.

International Conference of Building Officials

The International Conference of Building Officials was founded in 1922 for the purpose of the development of a code that all communities could accept and enforce. This goal was realized in 1927 with the publication of the Uniform Building Code. Immediately adopted by its supporters, use of the code has spread throughout the majority of the 50 states as well as the territories of many of the Pacific and Caribbean islands.

The International Conference of Building Officials is a non-profit service organization, owned and controlled by its member cities, counties and states. The Conference's aims and purposes are:

- Publication, maintenance and promotion of the Uniform Building Code and its related documents.

- Investigation and research of principles underlying safety to life and property in the construction, use and location of buildings and related structures.

- Development and promulgation of uniformity in regulations pertaining to building construction.

- Education of the building official.

- Formulation of guidelines for the administration of building inspection departments.

Reinforced Concrete Masonry Construction
Inspector's Handbook

SECOND EDITION

Introduction

1.1 GENERAL

This manual has been developed to provide the inspector with information and to serve as a general guide for **reinforced hollow concrete masonry construction.**

Reinforced hollow concrete masonry construction uses concrete blocks (also called concrete masonry units, or CMU for short) with steel reinforcing embedded in grout or mortar so that the separate materials acting together form a single structural system.

This publication has been prepared to assist masonry construction inspectors with the information needed to do a thorough professional job.

In order to understand a material and system, it is necessary to know its terminology. The first section of this book includes terms and definitions used in reinforced concrete masonry construction; Section 6 contains a more detailed glossary.

Since a construction project cannot begin until the proper materials are selected, materials is the next topic covered.

The Materials section is followed by Quality Control, Sampling and Testing, describing the sampling and testing of masonry necessary to assure that the materials used are in keeping with the prescribed standards and specifications.

Inspection of the actual construction is next, and this specifically deals with code concerns and inspection requirements of reinforced concrete masonry.

The handbook's last sections are on typical concrete masonry shapes, names and functions, glossary of terms, and Uniform Building Code Standards that relate to masonry.

1.2 THE INSPECTOR

A vital part of any construction project is good inspection. The inspector's job is important. Knowledge and good judgment are essential to obtaining the results required by the approved plans and specifications. The materials furnished on the job represent the manufacturers' efforts to supply products meeting job specifications. It is the inspector's responsibility to see that these products are properly used to produce a quality job.

1.3 RESPONSIBILITIES AND DUTIES

Prior to starting masonry construction, the inspector shall verify that necessary material testing has been performed as required. Some tests may have to be conducted well in advance of job site delivery such as for high strength block. All materials must meet specified requirements.

The inspector should keep a daily log from his first day on the project. The status of the job from the beginning should be noted.

The daily log should record weather, temperature, and job conditions. The inspector should record all materials, test specimens and job progress and note what work was accomplished and where it was done. This includes laying of masonry units and grout pours that are completed.

It is also suggested that the inspector note how many masons are on the job each day and the delivery of materials. Any special conditions, problems or adverse events that may take place should be noted.

If there are job conferences, a list of who attended, what was accomplished, and the decisions made should also be noted.

Complete and thorough job records are very important, and the inspector is invaluable in maintaining these.

1.4 EQUIPMENT AND MATERIALS FOR THE INSPECTOR

As with all competent and skilled professionals and craftsmen, construction inspectors must have their tools and materials to properly carry out their inspection duties and responsibilities. The following is a minimum suggested list that an inspector should have.

1. A current set of plans and specifications, including all change orders.

2. Applicable building codes and standards to which the project was designed and the requirements of the jurisdiction it is under.

3. A list of architects, engineers, contractors and subcontractors; names, addresses, telephone numbers and responsible persons.

4. A notebook or log to keep daily notes as described above.

5. Necessary forms for filing reports with required agencies.

6. Pens, pencils and erasers.

7. Folding rule or retractable tape and long steel tape.

8. String to check straightness.

9. Keel—yellow, blue and black.

10. Permanent felt tip markers for labeling specimens.

11. Hand level and plumb bob.

12. Small trowel and smooth rod for making and rodding mortar and grout samples.

13. Sample molds obtained from testing laboratory.

14. Absorbent paper towels and masking tape to take grout specimens.

There can be more items needed, depending on the project and scope of duties required of the inspector.

1.5 TERMINOLOGY

Masonry, like all materials, systems and specialties, has its own vocabulary. Knowing and understanding the terms is a basic requirement.

U.B.C. Sec. 2401(b) provides selected terms relative to masonry materials, design and construction with which masonry inspectors should be familiar.

U.B.C. Sec. 2401(b)

(b) **Definitions.** For the purpose of this chapter, certain terms are defined as follows:

AREAS:

Bedded area is the area of the surface of a masonry unit which is in contact with mortar in the plane of the joint.

Effective area of reinforcement (A_s) is the cross-sectional area of reinforcement multiplied by the cosine of the angle between the reinforcement and the direction for which effective area is to be determined.

Gross area is the total cross-sectional area of any plane encompassed by the outer periphery of any specified section.[See Figure 1.1.]

Net area is the gross cross-sectional area at any plane minus the area of ungrouted cores, notches, cells, unbedded areas, etc. Net area is the actual surface area of a cross section of masonry.[See Figure 1.2.]

Transformed area is the equivalent area of one material to a second based on the ratio of moduli of elasticity of the first material to the second.

BOND:

Adhesion bond is the adhesion between masonry units and mortar or grout.

Reinforcing bond is the adhesion between steel reinforcement and mortar or grout.

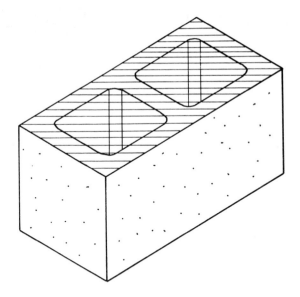

Figure 1.1 Gross area.

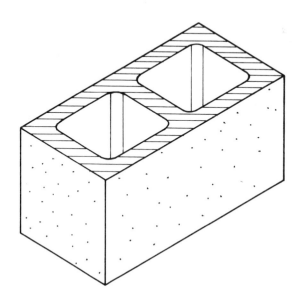

Figure 1.2 Net area.

CELL is a void space having a gross cross-sectional area greater than 1½ square inches.

CLEANOUT is an opening at the bottom of a grout space of sufficient size and spacing to allow the removal of debris. [See **Figure 1.3**.]

COLUMN is a vertical structural member whose horizontal dimension measured at right angles to the thickness does not exceed three times the thickness.

DIMENSIONS: [See **Figure 1.4**.]

Actual dimensions are the measured dimensions of a designated item; for example, a designated masonry unit or wall, as used in the structure. The actual dimension shall not vary from the specified dimension by more than the amount allowed in the appropriate standard of quality in Section 2402 of this chapter.

Nominal dimensions of masonry units are generally equal to its specified dimension plus the thickness of the joint with which the unit is to be laid.

Specified dimensions are the dimensions specified for the manufacture or construction of masonry, masonry units, joints or any other component of a structure. Unless otherwise stated, all calculations shall be made using or based on specified dimensions.

GROUT LIFT is an increment of grout height within the total pour; a pour may consist of one or more grout lifts.

GROUT POUR is the total height of masonry wall to be poured prior to the erection of additional masonry. A grout pour will consist of one or more grout lifts.

GROUTED MASONRY:

Grouted hollow unit masonry is that form of grouted masonry construction in which certain designated cells of hollow units are continuously filled with grout.

Grouted multiwythe masonry is that form of grouted masonry construction in which the space between the wythes is solidly or periodically filled with grout.

Figure 1.3 Cleanouts.

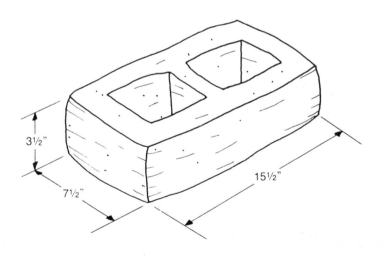

Figure 1.4 Nominal 8″ × 4″ × 16″ slumped concrete block, actual dimensions 7½″ × 3½″ × 15½″

JOINTS: [See Figure 1.5.]

Bed joint is the mortar joint that is horizontal at the time the masonry units are placed.

Collar joint is the vertical space separating a wythe of masonry from another wythe or from another continuous material and filled with mortar or grout.

Head joint is the mortar joint between units in the same wythe, usually vertical.

MASONRY UNIT is brick, tile, stone, glass block or concrete block conforming to the requirements specified in Section 2402.

Hollow masonry unit is a masonry unit whose net cross-sectional area in every plane parallel to the bearing surface is less than 75 percent of the gross cross-sectional area in the same plane.

Solid masonry unit is a masonry unit whose net cross-sectional area in every plane parallel to the bearing surface is 75 percent or more of the gross cross-sectional area in the same plane.

PRISM is an assemblage of masonry units, mortar and sometimes grout used as a test specimen for determining properties of the masonry.

REINFORCED MASONRY is that form of masonry construction in which reinforcement acting in conjunction with the masonry is used to resist forces and is designed in accordance with Section 2409.

SHELL is the outer portion of a hollow masonry unit as placed in masonry. [See Figure 1.5.]

WALL TIE is a mechanical fastener which connects wythes of masonry to each other or to other materials.

WALLS:

Bonded wall is a wall in which two or more of its wythes of masonry are adequately bonded together to act as a structural unit.

8

Cavity wall is a wall containing continuous air space with a minimum width of 2 inches and a maximum width of 4½ inches between wythes and the wythes are tied together with metal ties.

WEB is an interior solid portion of a hollow masonry unit as placed in masonry.

WYTHE is the portion of a wall which is one masonry unit in thickness. A collar joint is not considered a wythe.

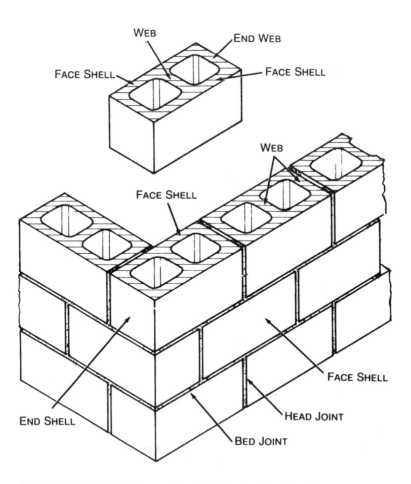

Figure 1.5 *Head joints and bed joints, shell and web.*

Materials

2.1 GENERAL

All materials used in reinforced concrete masonry construction shall conform to standard requirements.

2.2 MATERIAL STANDARDS

The standards for reinforced concrete masonry construction are set forth in the 1988 Uniform Building Code and Standards.

Relevant Uniform Building Code Standards are reproduced in Section 7.

UBC Sec. 2402

Material Standards

Sec. 2402. (a) Quality. Materials used in masonry shall conform to the requirements stated herein. If no requirements are specified in this section for a material, quality shall be based upon generally accepted good practice, subject to the approval of the building official.

(b) **Standards of Quality.**

1. **Aggregates:**
   A. U.B.C. Standard No. 24–21, Aggregate for Masonry Mortar.
   B. U.B.C. Standard No. 24–23, Aggregate for Masonry Grout.

2. **Cement:**
   A. U.B.C. Standard No. 24–16, Masonry Cement.
   B. U.B.C. Standard No. 26–1, Portland Cement.
   C. Plastic Cement: Plastic cement shall meet the requirements for portland cement as set forth in U.B.C. Standard No. 26–1 except in respect to limitations on insoluble residue, air entrainment, and additions subsequent to calcination. Approved types of plasticizing agents shall be added to portland cement Type I or II in the manufacturing process, but not in excess of 12 percent of the total volume.

3. **Lime:**
   A. U.B.C. Standard No. 24–17, Quick Lime for Structural Purposes.
   B. U.B.C. Standard No. 24–18, Hydrated Lime for Masonry Purposes.

4. **Masonry units—clay or shale:**
   A. U.B.C. Standard No. 24–8, Structural Clay Loadbearing Wall Tile.
   B. U.B.C. Standard No. 24–9, Structural Clay Non load-bearing Tile.
   C. U.B.C. Standard No. 24–1, Section 24.101, Building Brick (Solid Units).
   D. U.B.C. Standard No. 24–25, Ceramic Glazed Structural Clay Facing Tile, Facing Brick and Solid Masonry Units.
      (i) Load-bearing glazed brick shall conform to the weathering and structural requirements of U.B.C. Standard No. 24–1, Section 24.104, Facing Brick.
   E. U.B.C. Standard No. 24–8, Structural Clay Facing Tile.
   F. U.B.C. Standard No. 24–1, Section 24.104, Facing Brick (Solid Units).
   G. U.B.C. Standard No. 24–1, Section 24.105, Hollow Brick.

5. Masonry units—concrete:
A. U.B.C. Standard No. 24-3, Concrete Building Brick.
B. U.B.C. Standard No. 24-4, Hollow and Solid Load-bearing Concrete Masonry Units.
C. U.B.C. Standard No. 24-6, Non load-bearing Concrete Masonry Units.

6. Masonry units—other:
A. **Calcium silicate:**
      (i) U.B.C. Standard No. 24-2, Calcium Silicate Face Brick (Sand-Lime Brick).
B. **Glass block:**
      (i) Glass block may be solid or hollow and contain inserts.
      (ii) All mortar contact surfaces shall be treated to ensure adhesion between mortar and glass.
C. U.B.C. Standard No. 24-14, Unburned Clay Masonry Units.
D. U.B.C. Standard No. 24-13, Cast Stone.
E. **Reclaimed units:**
      (i) Reclaimed or previously used masonry units shall meet the applicable requirements as for new masonry units of the same material for their intended use.

7. Metal ties and anchors:
A. Metal ties and anchors shall be made of a material having a minimum tensile yield stress of 30,000 psi.
B. All such items not fully embedded in mortar or grout shall be coated with copper, cadmium, zinc or a metal having at least equivalent corrosion-resistant properties.

8. Mortar:
U.B.C. Standard No. 24-20, Mortar for Unit Masonry.

9. Grout:
U.B.C. Standard No. 24-29 Grout for Masonry and Section 2403 (d).

10. Reinforcement:
A. U.B.C. Standard No. 24-15, Part I, Joint Reinforcement for Masonry, and Part II Cold Drawn Steel Wire for Concrete Reinforcement.

B. U.B.C. Standard No. 26-4, Deformed and Plain Billet-Steel Bars for Concrete Reinforcement.
C. U.B.C. Standard No. 26-4, Rail-Steel Deformed and Plain Bars for Concrete Reinforcement.
D. U.B.C. Standard No. 26-4, Axle-Steel Deformed and Plain Bars for Concrete Reinforcement.
E. U.B.C. Standard No. 26-4, Deformed Low-Alloy Bars.

11. **Water.** Water used in mortar or grout shall be clean and free of deleterious amounts of acid, alkalies or organic material or other harmful substances.

2.3 CONCRETE MASONRY UNITS

2.3.1 General

The inspector's job site check of concrete masonry units should include visually inspecting for and rejecting broken or cracked units. Unless specifically noted in the specifications, minor cracks incidental to usual manufacturing, or minor chipping resulting from normal handling or shipping are not grounds for rejection. Inspection should also verify that color and texture comply with the approved sample when required.

As an additional check, the inspector may break a unit, note the proportion of broken aggregate showing on the fractured face, and look for internal evidence of moisture. If no aggregate is broken, the inspector may recheck to be sure that the units have been tested in the laboratory and meet all required specifictions. If moisture rings are apparent on the fractured face, he should recheck the age of the units and the laboratory tests for their moisture content.

2.3.2 Dimensions

Concrete masonry units (CMU) are commonly designated by their nominal dimensions, width, height and length (in that order), followed by a brief description, for example: 8'' x 4'' x 16'' split face.

Actual unit dimensions, such as 7-5/8'' x 3-5/8'' x 15-5/8'', are generally 3/8'' less than the nominal dimensions, which would be 8'' x 4'' x 16''. This allows for the typical 3/8'' mortar joint used in CMU construction and retains a modular dimension in increments of four inches. See **Figure 2.1**.

Slumped block unit dimensions, illustrated in **Figure 2.2**, are generally 1/2'' less and may vary depending on the characteristics of the particular units used.

2.3.3 Wide Selection of Units

There are also a large variety of specialty concrete masonry units that have been developed for special purposes. Specialty units have been developed for sound control, energy-efficient use of insulation, rapid placing mortarless block systems, paving blocks, pilaster units, and others. **Figure 2.3** shows some of these specialty units.

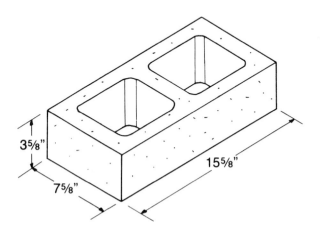

Figure 2.1 Precision concrete masonry unit.

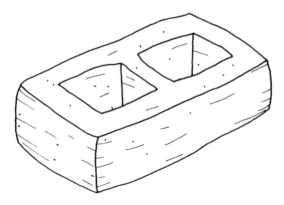

Figure 2.2
Slumped concrete masonry unit.

2.3.4 Component Units

Another type of specialty concrete masonry unit an inspector may encounter is known as a component system. Component systems provide added versatility for the designer and engineer by allowing the wall to be built to any desired thickness. Wall thicknesses are usually from eight inches to 24 inches in one inch increments. See **Figure 2.4**.

The system is used in retaining walls, subterranean walls, structural building walls, or as forms for concrete walls.

The masonry components are solid concrete blocks conforming to ASTM C55 with a 2500 psi compressive strength. An example is shown in **Figure 2.5**.

The components can be assembled with different architectural finishes on each side. They may also be used as permanent forms for poured-in-place concrete.

2.3.5 Storing Masonry Units

Care must be taken when storing concrete masonry units on the job site to ensure they are clean and dry when used, as shown in **Figure 2.6**. Concrete masonry units shall not be wetted unless otherwise approved.

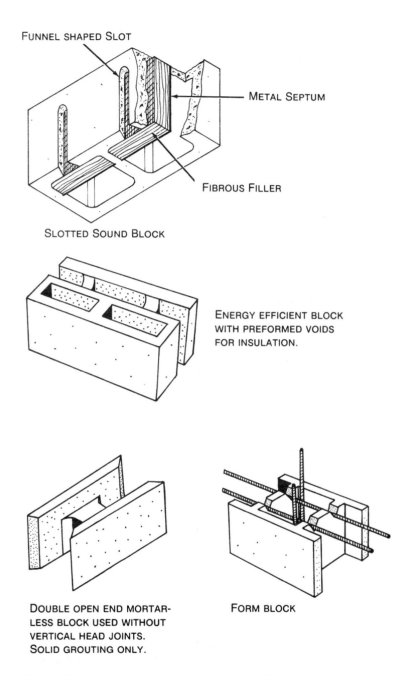

FUNNEL SHAPED SLOT

METAL SEPTUM

FIBROUS FILLER

SLOTTED SOUND BLOCK

ENERGY EFFICIENT BLOCK
WITH PREFORMED VOIDS
FOR INSULATION.

DOUBLE OPEN END MORTAR-
LESS BLOCK USED WITHOUT
VERTICAL HEAD JOINTS.
SOLID GROUTING ONLY.

FORM BLOCK

Figure 2.3 *Specialty concrete masonry units.*

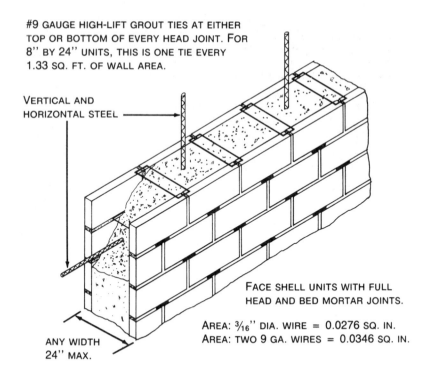

#9 GAUGE HIGH-LIFT GROUT TIES AT EITHER TOP OR BOTTOM OF EVERY HEAD JOINT. FOR 8" BY 24" UNITS, THIS IS ONE TIE EVERY 1.33 SQ. FT. OF WALL AREA.

VERTICAL AND HORIZONTAL STEEL

FACE SHELL UNITS WITH FULL HEAD AND BED MORTAR JOINTS.

AREA: 3/16" DIA. WIRE = 0.0276 SQ. IN.
AREA: TWO 9 GA. WIRES = 0.0346 SQ. IN.

ANY WIDTH 24" MAX.

Figure 2.4 *Expandable component masonry system.*

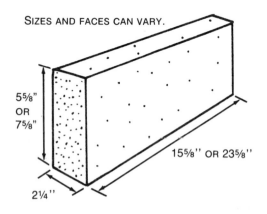

SIZES AND FACES CAN VARY.

5⅝" OR 7⅝"

15⅝" OR 23⅝"

2¼"

Figure 2.5 *Component unit.*

Figure 2.6 *Properly stored masonry units.*

2.4 CEMENTITIOUS MATERIALS

For masonry units to function effectively in a wall, it is important that they be bonded together. This bonding is achieved with mortar and grout. The adhesion is obtained with the cementitious materials cement and lime.

Attempts have been made to increase the economy of block masonry by developing mortarless block systems that can interlock and be laid without mortar. Another approach has been to manufacture the units with very uniform bearing surfaces, often achieved by grinding the edges. The blocks are then laid with very thin high bond mortars.

2.4.1 Portland Cement

The classical definition of portland cement is: ''Portland cement is the product obtained by finely pulverizing a clinker made by calcining to incipient fusion an intimate and properly proportioned mixture of argillaceous and calcareous material with no additions subsequent to calcination except water and calcined or uncalcined gypsum.''

Portland cement is the primary bonding agent used to glue together the grains of sand and pea gravel used in mortar and grout.

Portland cement must conform to the requirements of Uniform Building Code Standard No. 26-1 and the ASTM C150-81.

Portland cement must be properly stored off the ground and covered to prevent absorption of moisture. When the mortar joint color is critical, the same type and brand of cement and lime and aggregate should be used throughout the job. Sacks with hard lumps should be rejected. Usually Type I or Type II portland cement is used for mortar and grout. In some instances, low alkali portland cement, if available, may be used to reduce the possibility of efflorescence.

2.4.2 Plastic Cement

In some of the southwestern areas of the United States, plastic cement is sometimes used for mortar. This is basically Type I portland cement with approximately 12 percent plasticizing agent added. When plastic portland cement is used in mortar, hydrated lime may be added, but not in excess of one-tenth of the volume of cement.

Plastic cement is generally used for small masonry projects and the "do it yourself" home masonry market since lime does not have to be used to obtain adequate plasticity. Mortar made with 1 part plastic cement and 3 parts sand is equivalent to a mix of 1 part portland cement, 0.14 parts plasticizer and 3.4 parts sand which is richer than type S, portland cement, lime mortar.

2.4.3 Mortar Cement

In some parts of the United States, portland cement manufacturers and some masonry materials supply companies may package a blend of portland cement and hydrated lime. This blending may be in proportions by volume for Type M mortar, one part portland cement and 1/4 part lime; for Type S mortar, one part portland cement and 1/2 part lime; and for Type N mortar, one part portland cement and one part lime, or prepackaged to specifications.

This packaged mortar cement conforms to the requirements of Table 24-A of the Uniform Building Code. It is generally used for small projects or jobs where separate delivery of portland cement and lime is inconvenient.

2.4.4 Masonry Cement

Masonry cement is a mixture of portland cement, 30% to 60% plasticizer material, and added chemicals. The Uniform Building Code Standard No. 24-16 for "Cement, Masonry" is based on ASTM C91-67. This 1967 standard is for a masonry cement that is satisfactory for Type N mortar. If it is to be used for Type M or Type S mortar, additional portland cement must be used in the mortar mix.

The ASTM standard specification C91 "Masonry Cement" has been revised and was reissued in 1983. This 1983 ASTM standard specification covers three types of masonry cement for use in mortar. These are Type M, Type S and Type N masonry cements that may be used for mortar without the addition of more portland cement. The particular types of masonry cements are blended to produce mortar of the same type to conform to ASTM C270-84 "Mortar for Unit Masonry."

2.4.5 Lime

The Uniform Building Code permits the use of hydrated lime or lime putty in mortar. The use of lime putty has given way to convenient packaged hydrated lime that is delivered in sacks.

Hydrated limes are divided into two classes, as descibed in ASTM C207, which are Type N and Type S hydrated limes. These are high calcium and dolomitic, high magnesium, hydrates. The Type S, special hydrated lime, is different from Type N, normal hydrated lime, principally by its ability to develop high early plasticity, higher water retentivity, and by its limitation on unhydrated oxide content.

Lime use in mortar improves the plasticity of the mix; it improves water retention for longer board life; it improves the water-tightness of the mortar joint, increases the bond between the mortar and the masonry unit, and it contributes to the cementitious materials in the mortar mix.

Increasing the portland cement content and reducing the lime content increases the compressive strength of mortar, but it also increases shrinkage, reduces workability, lowers water retentivity and causes rapid stiffening.

Conversely, increasing the lime improves workability, water retentivity and adhesion bond; it does not add to the compressive strength of mortar but it does aid waterproofing of the mortar. **Figure 2.7** shows the relationship between various proportions of cement and lime and mortar strength and water retentivity.

The Uniform Building Code Table No. 24-B "Grout Proportions by Volume" allows up to one-tenth parts by volume hydrated lime. This allowance is believed to be a carry-over from when mortar was used as a slushing grout material. Although lime is not generally

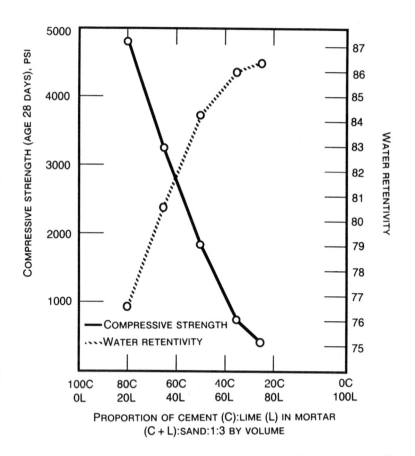

Figure 2.7 *Relation between mortar composition, compressive strength, and water retentivity. (Ref. No. 14)*

used in grout, it may occasionally be used as a lubricant to initially charge grout pumps.

2.5 AGGREGATES FOR MORTAR AND GROUT

Aggregates for mortar and grout are composed of sand and pea gravel.

Aggregates should be stored in a level, dry, clean place from which they can be measured into the mixer with minimum handling and kept free from contamination by harmful substances.

Aggregates should be delivered to the job pregraded and the gradation certified by the supplier. The inspector need only check the certificate and observe the aggregate for consistent gradations. Field tests will need to be made when called for by the specifications. Field tests are generally sieve analysis tests.

2.6 REINFORCING STEEL

2.6.1 General

Reinforcing steel at the job site must be protected from accidental kinking or bending. It must also be kept free of dirt, mud, oil or other foreign matter detrimental to bond. Light surface rust or light mill scale is not detrimental to bond provided the unit weight after the specimen has been cleaned still meets minimum ASTM weight and height of deformation requirements.

Reinforcing steel must be placed as detailed in the plans and specifications. If, for any reason, the reinforcement cannot be placed as designed, the architect and/or engineer should be notified prior to construction.

The inspector shall check the reinforcing bars to assure that they are the grade and size specified. **Figure 2.8** shows the markings for identifiction of reinforcing bars. **Table 2-A** and **Table 2-B** provide information on the properties of reinforcing bars.

2.6.2 Reinforcing Bars

In the Western United States, and particularly in California, the majority of reinforcing steel used in masonry is deformed bars. The deformed bars range from #3 (3/8" in diameter) to a recommended

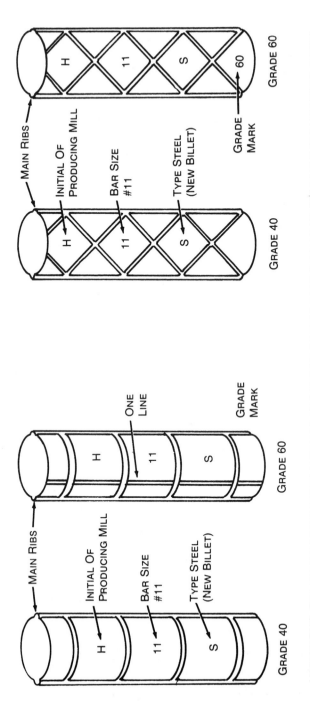

VARIATIONS: BAR IDENTIFICATION MARKS MAY ALSO BE ORIENTED TO READ HORIZONTALLY (AT 90° TO THOSE ILLUSTRATED ABOVE).

GRADE MARK LINES MUST BE CONTINUED AT LEAST FIVE DEFORMATION SPACES.

GRADE MARK NUMBERS MAY BE PLACED WITHIN SEPARATE CONSECUTIVE DEFORMATION SPACES TO READ VERTICALLY OR HORIZONTALLY.

Figure 2.8 *Identification marks, line system of grade marks.*

24

Table 2-A Properties of Standard ASTM A615 Steel Reinforcing Bars

Bar Designation Number[b]	Nominal Weight, lbs. per ft.	Nominal Dimensions[a]			Deformation Requirements, In.		
		Diameter in.	Cross-Sectional Area, sq. in.	Perimeter in.	Maximum Average Spacing	Minimum Average Height	Maximum Gap (Chord of 12½ Percent of Nominal Perimeter)
3	0.376	0.375	0.11	1.178	0.262	0.015	0.143
4	0.668	0.500	0.20	1.571	0.350	0.020	0.191
5	1.043	0.625	0.31	1.963	0.437	0.028	0.239
6	1.502	0.750	0.44	2.456	0.525	0.038	0.286
7	2.044	0.875	0.60	2.749	0.612	0.044	0.334
8	2.670	1.000	0.79	3.142	0.700	0.050	0.383
9	3.400	1.128	1.00	3.544	0.790	0.056	0.431
10	4.303	1.270	1.27	3.990	0.889	0.064	0.487
11	5.313	1.410	1.56	4.430	0.987	0.071	0.540
14[c]	7.650	1.693	2.25	5.320	1.185	0.085	0.648
18[c]	13.600	2.257	4.00	7.090	1.580	0.102	0.864

[a]The nominal dimensions of a deformed bar are equivalent to those of a plain round bar having the same weight per foot as the deformed bar.

[b]Bar numbers are based on the number of eighths of an inch included in the nominal diameter of the bars.

[c]Not permitted in masonry construction. Maximum size bars for masonry are No. 11.

Table 2-B *Overall Diameter of Bars*

Bar Size	Approx. dia. outside Deformations (inches)
# 3	7/16
# 4	9/16
# 5	11/16
# 6	7/8
# 7	1
# 8	1-1/8
# 9	1-1/4
#10	1-7/16
#11	1-5/8
#14	1-7/8
#18	2-1/2

SECTION A-A

OVERALL DIAMETER

NOTE that diameters tabulated are approximate size outside of deformations, so clearance should be added for holes.

maximum of #11 bars (1-3/8'' in diameter). This reinforcing steel conforms to ASTM A615-82, which specifies the physical characteristics of the reinforcing steel. Reinforcing steel may be either Grade 40 with a minimum yield strength of 40,000 psi or Grade 60 with a minimum yield strength of 60,000 psi.

Grade 40 steel bars are furnished in sizes 3, 4, 5 and 6. However, currently, Grade 60 steel is furnished in all sizes, and if Grade 40 is required, special note must be made to ensure delivery.

2.6.3 Identification Marks

The ASTM specifications covering new billet steel, rail steel and axle steel reinforcing bars (A615, A616, A617 and A706) require identification marks to be rolled into the surface of one side of the bar to denote the producer's mill designation, bar size, and type of

Materials

steel, and for Grade 60, grade marks indicating yield strength. See **Figure 2.8**, Identification marks, line system of grade marks.

Grade 40 bars show only three marks (no grade mark) in the following order:

1st — Producing Mill (usually an initial)

2nd — Bar Size Number (#3 through #18)

3rd — Type S for New Billet, A for Axle, I for Rail, W for Low Alloy

Grade 60 bars must also show grade marks:
60 or One (1) Line for 60,000 psi strength

Grade mark lines are smaller and between the two main longitudinal ribs which are on opposite sides of all U.S.-made bars. Number grade marks are fourth in order.

2.6.4 Overall Bar Diameters

Bar diameters are nominal, with the actual diameter outside of deformations being somewhat greater. The outside diameter may be important when punching holes in structural steel members to accommodate bars or when allowing for the out-to-out width of a group of beam bars crossing and in contact with column verticals. Approximately 1/16 in. for #3, #4, #5 bars, 1/8 in. for #6, #7, #8, #9 bars, 3/16 in. for #10 and #11 bars should be added to the nominal bar diameter for the height of the deformations. See **Table 2-B**.

2.7 JOINT REINFORCING STEEL

2.7.1 General

When high strength steel wire, in ladder or truss type configuration, is placed in the horizontal bed joints, it is called joint reinforcing.

High strength steel wire fabricated in ladder or truss systems, as illustrated in **Figure 2.9**, is placed in the bed joints to reinforce the wall in the horizontal direction. The most common uses of joint reinforcing are (1) to control shrinkage cracking in concrete masonry walls; (2) as part or all of the "minimum" steel required by the Uniform Building Code; and (3) as designed reinforcing that resists

27

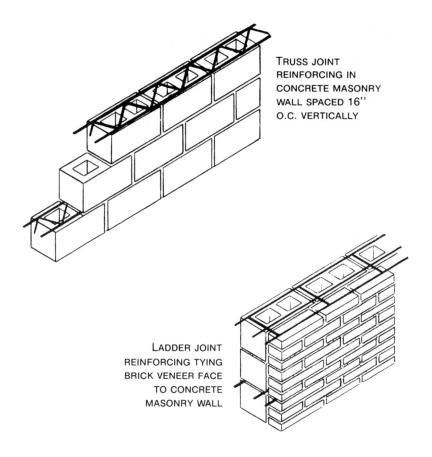

TRUSS JOINT
REINFORCING IN
CONCRETE MASONRY
WALL SPACED 16"
O.C. VERTICALLY

LADDER JOINT
REINFORCING TYING
BRICK VENEER FACE
TO CONCRETE
MASONRY WALL

Figure 2.9 *Use of joint reinforcing.*

forces in the masonry, such as tension and shear. It can also be used as reinforcing in all types of masonry walls as a continuous tie system for veneer and cavity walls.

2.7.2 Description

Joint reinforcement consists of deformed longitudinal wires welded to cross wires in sizes suitable for placing in mortar joints between masonry courses. [See **Figure 2.9**]

2.7.3 Configuration and Size of Longitudinal and Cross Wires

(a) **General.** The distance between longitudinal wires and the configuration of cross wires connecting the longitudinal wires shall conform to the design.

(b) **Longitudinal Wires.** The diameter of longitudinal wires shall not be less than 0.148 inch (No. 9 gauge) nor more than one-half the mortar joint thickness.

(c) **Cross Wires.** The diameter of cross wires shall not be less than (No. 9 gauge) 0.148 inch diameter nor more than the diameter of the longitudinal wires. Cross wires shall not project beyond the outside longitudinal wires by more than ⅛ inch.

(d) **Width.** The width of joint reinforcement shall be the out-to-out distance between outside longitudinal wires. Variation in the width shall not exceed ⅛ inch.

(e) **Length.** The length of pieces of joint reinforcment shall not vary more than ½ inch or 1.0 percent of the specified length, whichever is less.

2.7.4 Material Requirements

(a) **Tensile Properties.** Wire of the finished product shall meet the following requirements:

Tensile strength, minimum	75,000 psi
Yield strength, minimum	60,000 psi
Reduction of area, minimum	30 percent

For wire testing over 100,000 psi, the reduction of area shall not be less than 25 percent.

(b) **Bend Properties.** Wire shall not break or crack along the outside diameter of the bend when tested in accordance with Section 24.1508.

(c) **Weld Shear Properties.** The least weld shear strength in pounds shall be not less than 25,000 multiplied by the specified area of the smaller wire in square inches.

2.7.5 Fabrication

Wire shall be fabricated and finished in a workmanlike manner, shall be free from injurious imperfections and shall conform to this standard.

The wires shall be assembled by automatic machines or by other suitable mechanical means which will assure accurate spacing and alignment of all members of the finished product.

Longitudinal and cross wires shall be securely connected at every intersection by a process of electric-resistance welding.

Longitudinal wires shall be deformed. One set of four deformations shall occur around the perimeter of the wire at a maximum spacing of 0.7 times the diameter of the wire but not less than eight sets per inch of length. The overall length of each deformation within the set shall be such that the summation of gaps between the ends of the deformations shall not exceed 25 percent of the perimeter of the wire. The height or depth of the deformations shall be 0.012 inch for 3/16 inch diameter or larger wire, 0.011 for No. 8 gauge wire (0.162-inch diameter) and 0.009 inch for No. 9 gauge wire (0.148-inch diameter).

2.8 WATER

Water used in masonry construction should be potable (suitable for drinking) and free of harmful substances such as oil, acids, alkalis, etc.

2.9 ADDITIVES AND ADMIXTURES

Sometimes it is desirable to have certain qualities, such as delayed setting, super plasticity, water reduction, accelerated strength gain, etc. These special qualities can be obtained, sometimes, by using special additives or admixtures in the mortar or grout. To use the additives, it is important to comply with the recommendations of the manufacturer to obtain satisfactory results after first obtaining approval of the local building official.

The Uniform Building Code Sec. 2403(e) provides general requirements.

U.B.C. Sec. 2403(e)

(e) **Additives and Admixtures. 1. General.** Additives and admixtures to mortar or grout shall not be used unless approved by the building official.

2. Antifreeze compounds. Antifreeze liquids, chloride salts or other substances shall not be used in mortar or grout.

3. Air entrainment. Air-entraining substances shall not be used in mortar or grout unless tests are conducted to determine compliance with the requirements of this code.

2.10 MORTAR

2.10.1 General

Mortar is a basic component of reinforced masonry. Some claim that mortar holds the units apart, others claim it holds the masonry units together, it actually does both.

Mortar has been made from many different materials. Some ancient mortar mixtures were plain mud or clay, earth with ashes, ox blood and earth, and sand and lime.

Modern mortar consists of cementitious materials and well graded sand with sufficient fines. Mortar is used for the following purposes:

a. It is a bedding or seating material for the masonry unit.

b. It allows the unit to be leveled and properly placed.

c. It bonds the units together.

d. It provides compressive strength.

e. It provides shear strength, particularly parallel to the wall.

f. It allows some movement and elasticity between units.

g. It seals irregularities of the masonry unit and provides a weather-tight wall, prevents penetration of wind and water into and through the wall.

h. It can provide color to the wall by using mineral color additive.

i. It can provide an architectural appearance by using various types of joints, as shown in **Figures 4.24** and **4.25**.

Section 2403 of the Uniform Building Code states the requirements for mortar.

U.B.C. Sec. 2403 (a), (b), (c).1.

Mortar and Grout

Sec. 2403. (a) **General.** Mortar and grout shall comply with the provisions of this section. Special mortars, grouts or bonding systems may be used, subject to satisfactory evidence of their capabilities when approved by the building official.

(b) **Materials.** Materials used as ingredients in mortar and grout shall conform to the applicable requirements given in Section 2402. Cementitious materials shall be lime, masonry cement and portland cement.

(c) **Mortar.** 1. **General.** Mortar shall consist of a mixture of cementitious material and aggregate to which sufficient water and approved additives, if any, have been added to achieve a workable, plastic consistency.

2.10.2 Proportions of Mortar

Proportions of mortar may be based on laboratory testing based on prisms, cube strength, and cylinder strengths. Field experience based on history of performance with the mortar ingredients and masonry units for the project may be used as a basis for proportions.

Long-time experience has proved that mortar portions based on U.B.C. Table No. 24-A Mortar Proportions by Volume for Unit Masonry results in satisfactory performance.

U.B.C. Sec. 2403 (a)2.

2. **Selecting proportions.** Proportions of ingredients and any additives shall be based on laboratory or field experience with the mortar ingredients and the masonry units to be used. The mortar shall be specified by the proportions of its constituents in terms of parts by volume. Water content shall be adjusted to provide proper workability under existing field conditions. When the proportion of ingredients is not specified, the proportions by mortar type shall be used as given in Table No. 24-A.

TABLE NO. 24-A—MORTAR PROPORTIONS FOR UNIT MASONRY

MORTAR	TYPE	PORTLAND CEMENT OR BLENDED CEMENT[1]	MASONRY CEMENT[2] M	MASONRY CEMENT[2] S	MASONRY CEMENT[2] N	HYDRATED LIME OR LIME PUTTY[1]	AGGREGATE MEASURED IN A DAMP, LOOSE CONDITION
Cement-lime	M	1	—	—	—	¼	Not less than 2¼ and not more than 3 times the sum of the separate volumes of cementitious materials.
	S	1	—	—	—	over ¼ to ½	
	N	1	—	—	—	over ½ to 1¼	
	O	1	—	—	—	over 1¼ to 2½	
Masonry cement	M	1	—	—	1	—	
	M	—	1	—	—	—	
	S	½	—	—	1	—	
	S	—	—	1	—	—	
	N	—	—	—	1	—	
	O	—	—	—	1	—	

[1] When plastic cement is used in lieu of portland cement, hydrated lime or putty may be added, but not in excess of one tenth of the volume of cement.
[2] Masonry cement conforming to the requirements of U.B.C. Standard No. 24–16.

In Seismic Zones 3 and 4, only Type M and Type S mortar can be used for the structural system. In Seismic Zones 0, 1 and 2, Types M, S and N can be used for the structural system.

The use of masonry cement and plastic cement is prohibited in Seismic Zones 2, 3 and 4 for the structural system.

Using masonry cement that conforms to the requirements of U.B.C. Standard No. 24–16 or ASTM C91-83, Types M, S or N mortar can be obtained without the addition of extra portland cement, as given in **Table 2-C**.

Field practice is to use the range of proportions for each type of mortar that will result in a workable, smooth mortar that spreads easily and is plastic enough to be able to push the masonry unit into it. It must also be stiff enough to support the masonry unit without deforming under the additional weight of masonry units.

Table 2-C *Mortar Proportions by Volume for Masonry*

Mortar Type	Parts By Volume of Portland Cement[1]	Parts By Volume of Masonry Cement[2]	Parts By Volume of Hydrated Lime or Lime Putty[1]	Aggregate Measured in a Damp, Loose Condition
M	1	1(N)	–	2-1/4 to 3 times the sum of the cementitious materials, Portland Cement, masonry cement and lime.
	1	–	1/4	
	–	1(M)	–	
S	1/2	1(N)	–	
	1	–	over 1/4 to 1/2	
	–	1(S)	–	
N	–	1(N)	–	
	1	–	over 1/2 to 1-1/4	
O	–	1(N)	–	
	1	–	over 1-1/4 to 2-1/2	

[1]When plastic cement is used in lieu of portland cement, hydrated lime or putty may be added, but not in excess of one tenth of the volume of cement.

[2]Type of masonry cement indicated in parenthesis, conforming to U.B.C. Standard No. 24–16.

Type S mortar, made with portland cement and hydrated lime, can be proportioned with one part portland cement, one quarter to one-half part hydrated lime and 2-3/4 to 4-1/2 parts sand. The variation in sand proportions allows an adjustment due to particle shape, size, and grading, all of which affect workability and spreadability.

2.10.3 Mortar Aggregate—Sand

The aggregate used for mortar should be well graded with sufficient fine material passing the No. 100 sieve to impart smoothness to the mortar. Plaster sand is ideal for mortar for it has no particle larger than 1/8'' and it has sufficient fines for workability and smoothness.

Particle shape influences the workability of mortar. Round, spherical particles, well graded, are best for mortar while sharp, cubical or flat particles produce harsh mortar.

U.B.C. STD. TABLE NO. 24–21–A—AGGREGATE FOR MASONRY MORTAR

SIEVE SIZE	PERCENT PASSING	
	NATURAL SAND	MANUFACTURED SAND
No. 4 (4750-micron)	100	100
No. 8 (2360-micron)	95 to 100	95 to 100
No. 16 (1180-micron)	70 to 100	70 to 100
No. 30 (600-micron)	40 to 75	40 to 75
No. 50 (300-micron)	10 to 35	20 to 40
No. 100 (150-micron)	2 to 15	10 to 25
No. 200 (75-micron)	—	0 to 10

The aggregate shall not have more than 50 percent retained between any two consecutive sieves shown in the table above. There shall be not more than 25 percent retained between the No. 50 and the No. 100 sieve.

The fineness modulus shall not vary by more than 0.20 from the value assumed in selecting proportions for the mortar, unless suitable adjustments are made in proportions to compensate for the change in grading.

Concrete sand should not be used because the maximum grain sizes may be 3/16'' to 1/4'' and needed fines have been washed out resulting in a sand too harsh and coarse and unsuitable for mortar.

Aggregates should be stored in a level, dry, clean place from which they can be measured into the mixer with minimum handling and kept free from contamination by harmful substances.

Uniform Building Code Standard Table No. 24-21-A Aggregate for Masonry Mortar gives the grading requirements for sand.

2.10.4 Mixing

Mortar mixing is best accomplished in a paddle type mixer. About one-half of the water and one quarter of the sand are put into the operating mixer first, then the cement, lime, color (if any), and the remaining water and sand are added. All materials should then mix for not less than three minutes and not more than ten minutes

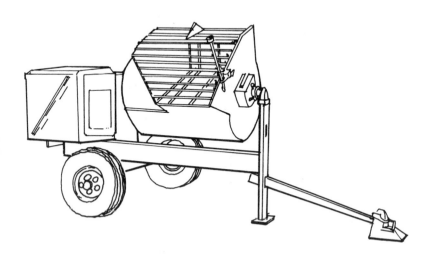

Figure 2.10 Plaster or paddle mortar mixer.

in a mechanical mixer with the amount of water required to provide the desired workability. Small amounts of mortar can be hand mixed.

In a paddle mixer, shown in **Figure 2.10**, the drum is stationary and the blades rotate through the mortar materials for thorough mixing.

A drum or barrel mixer, shown in **Figure 2.11**, rotates the drum in which the materials are placed. The material is carried to the top of the rotation and drops down to achieve mixing.

2.10.5 Retempering

Mortar may be retempered one time with water when needed to maintain workability. This should be done on mortar boards by forming a basin or hollow in the mortar, adding water, and then reworking the mortar into the water. Splashing water over the top of the

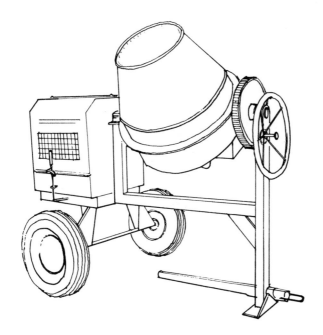

Figure 2.11 Drum or barrel concrete mixer.

mortar is not permissible. Harsh mortar, mortar that has begun to stiffen or harden due to hydration, should be thrown out. Mortar must be used within two-and-one-half hours after the initial water has been added to the dry ingredients at the job site.

2.10.6 Color

Mortar colors are generally mineral oxides or carbon black. Iron oxide is used for red, yellow, and brown colors; chromium oxide is for green, and cobalt oxide is for blue colors.

The amount of color additive depends on the color and intensity and ranges from 0.5% to 7.0% for the mineral oxides and a maximum of 3% for carbon black. The percent is based on weight of cement content. These maximum percentages are far greater than the normal amounts of color added.

There are commercially prepared colors for mortars that offer a wide variety of colors and shades.

U.B.C. Sec. 2403(e)4.

4. **Colors.** Only pure mineral oxide, carbon black or synthetic colors may be used. Carbon black shall be limited to a maximum of 3 percent of the weight of the cement.

Mixing time should be long enough for a uniform, even color to be obtained in the mortar and should be the same length of time for every batch.

Mixing sequence should be the same for each batch and as specified in Section 2.10.4 "Mixing."

Retempering must be kept to a minimum when coloring is used, and for best results should not be done at all.

Materials—Source and manufacturer and amount of each ingredient should remain the same for all colored mortar on the project so as to obtain the same color throughout. Prepackaged mineral color additives that can be added to the mix based on full sacks of portland cement provide a consistent batching for quality control of mortar color.

2.10.7 Proprietary Mortars

Proprietary mortars such as masonry cement mortars, delayed set mortars, and ready mix mortars, shall be approved by the engineer or architect and accepted by the building official. Handling and use of these shall be in strict compliance with the manufacturer's recommendations.

2.10.8 Mortar Admixtures

There are retarding admixtures that delay the set and stiffening of mortar. Retardation can be obtained for 36 hours or more.

There are also admixtures that are used to replace lime. These admixtures usually add air to the mortar mix to provide workability.

Admixtures shall be approved by the architect or engineer and be acceptable to the building official.

2.11 GROUT

2.11.1 General

Grout is a fluid mixture of cement, sand, and sometimes pea gravel; it is very plastic concrete with a slump of eight to ten inches. This high slump is necessary for the grout to flow into all the grout spaces and joints and completely surround the steel. The excess water is absorbed into the masonry units, thereby reducing the water/cement ratio of the grout. The absorbed water in the concrete masonry units aids in curing the grout and increasing the strength gain.

> **U.B.C. Sec. 2403(d)**
>
> (d) **Grout. 1. General.** Grout shall consist of a mixture of cementitious materials and aggregate to which water has been added such that the mixture will flow without segregation of the constituents.

2.11.2   Types of Grout

Fine Grout. Fine grout may be used in grout spaces in multi-wythe masonry as small as 3/4'' or larger and in grout spaces in hollow unit construction 1-1/2'' by 2'' or more in horizontal dimensions.

Coarse Grout. Coarse grout which uses pea gravel may be used in grout spaces in multi-wythe masonry 1-1/2'' or more in horizontal dimension and in grout spaces in hollow unit construction 1-1/2'' by 3'' or more in horizontal dimensions.

2.11.3   Proportions

Grout may be proportioned in accordance with U.B.C. Table No. 24-B or may be proportioned by laboratory designed mixes based on testing or field experience. The testing values would be based on masonry prism tests or grout specimen tests made in accordance with U.B.C. Standard No. 24-29, or field experience based on a history of performance with the same masonry units and grout materials and mix proportions used for the project. The use of 70% sand and 30% pea gravel requires six sacks of portland cement per cubic yard and results in a pumpable grout that will provide the strength required by U.B.C. Standard No. 24-29. Grout must have adequate strength for satisfactory f'_m values, for bonding reinforcing steel, for embeddment of anchor bolts, and for proper bond to transfer the stress to the reinforcing bar.

Long-time experience has proved that grout proportions based on U.B.C. Table No. 24-B and Sec. 2403(d)2 are successful for regular load-bearing concrete masonry.

U.B.C. TABLE NO. 24–B—GROUT PROPORTIONS BY VOLUME[1]

TYPE	PARTS BY VOLUME OF PORTLAND CEMENT OR BLENDED CEMENT	PARTS BY VOLUME OF HYDRATED LIME OR LIME PUTTY	AGGREGATE MEASURED IN A DAMP, LOOSE CONDITION	
			Fine	Coarse
Fine grout	1	0 to $\frac{1}{10}$	$2\frac{1}{4}$ to 3 times the sum of the volumes of the cementitious materials	
Coarse grout	1	0 to $\frac{1}{10}$	$2\frac{1}{4}$ to 3 times the sum of the volumes of the cementitious materials	1 to 2 times the sum of the volumes of the cementitious materials

[1] Grout shall attain a minimum compressive strength at 28 days of 2,000 psi. The building official may require a compressive field strength test of grout made in accordance with the U.B.C. Standard No. 24-22.

U.B.C. Sec. 2403(d)2.

2. **Selecting proportions.** Proportions of ingredients and any additives shall be based on laboratory or field experience with the grout ingredients and the masonry units to be used. The grout shall be specified by the proportion of its constituents in terms of parts by volume. Water content shall be adjusted to provide proper workability and to enable proper placement under existing field conditions, without segregation. When the proportion of ingredients is not specified, the proportions by grout type shall be used as given in Table No. 24–B or a minimum compressive strength shall be specified of at least 2000 psi.

2.11.4 Aggregate for Grout

Aggregate for grout shall conform to U.B.C. Standard No. 24-23. Grading shall be given in U.B.C Standard Table No. 24-23-A.

U.B.C. STD. TABLE NO. 24-23-A—GRADING REQUIREMENTS

SIEVE SIZE	AMOUNTS FINER THAN EACH LABORATORY SIEVE (Square Openings), Percent by Weight				
	Fine Aggregate			Coarse Aggregate	
	Size No. 1	Size No. 2			
		Natural	Manufac-tured	Size No. 8	Size No. 89
½-inch	—	—	—	100	100
⅜-inch	100	—	—	85 to 100	90 to 100
No. 4 (4.76-mm)	95 to 100	100	100	10 to 30	20 to 55
No. 8 (2.38-mm)	80 to 100	95 to 100	95 to 100	0 to 10	5 to 30
No. 16 (1.19-mm)	50 to 85	60 to 100	60 to 100	0 to 5	0 to 10
No. 30 (595-micron)	25 to 60	35 to 70	35 to 70	—	0 to 5
No. 50 (297-micron)	10 to 30	15 to 35	20 to 40	—	—
No. 100 (149-micron)	2 to 10	2 to 15	10 to 25	—	—
No. 200 (74-micron)	—	—	0 to 10	—	—

2.11.5 Mixing

Grout shall be mixed at the job site for at least three minutes to assure thorough blending of all ingredients, but not more than 10 minutes. Enough water must be used in the mixing process to achieve a high slump of eight to ten inches. This high slump is necessary for the grout to flow into the relatively small cells of the concrete masonry. Excess water is absorbed into the masonry, where it aids in the curing process.

2.11.6 Grout Admixtures

Admixtures used in grout impart desired properties. When admixtures are used, they should be approved by the architect or the engineer and the building official. Three admixtures used are:

a. shrinkage compensating admixtures to counteract the loss of water and the shrinkage of portland cement by creating an expansive gas in the grout;

b. super plasticizer admixtures to obtain high slump with reduced water in the grout. Grout with a 4'' slump will usually go to a 10'' slump with the use of a super plasticizer.

c. cement replacement such as fly ash is used in grout. Current provisions allow a maximum of 15% to 20% by weight of portland cement to be replaced by fly ash. The maximum amount is dependent on the fly ash, portland cement and strength gain characteristics.

Fly ash is a pozzolanic material obtained from combustion of coal which is collected in electrostatic precipitators or bag houses. It is classified by precise particle size and by chemical composition as Type F or C.

Type F fly ash is obtained from the combustion of anthracite, bituminous or sub-bituminous coal. It is low in lime, less than 7%, and contains greater than 70% silica, alumina and iron.

Type C fly ash comes from burning lignite or sub-bituminous coal and has more than 15% lime.

Fly ash, types F and C, are siliceous or siliceous and aluminous material which in itself possesses little or no cementitious value but will, in finely divided form and in the presence of moisture, react with calcium hydroxide at ordinary temperatures to form compounds possessing cementitious properties.

Fly ash, because of its fine, spherical particles, increases workability and cohesiveness. It reduces water demand and improves pumpability of grout.

When fly ash is used as a cement replacement, it is necessary to have assurance that the required strength will be obtained in the stated period of time.

2.11.7 Anti-Freeze Compounds

Most antifreeze admixtures are actually accelerators that increase the temperature by speeding up the hydration process.

Some antifreeze admixtures use alcohol to lower the freezing point; however, it requires a significant amount and this will reduce both the compressive strength and bond strength of the mortar.

U.B.C. Sec. 2403(e)2.

2. **Antifreeze compounds.** Antifreeze liquids, chloride salts or other such substances shall not be used in mortar or grout.

Quality Control, Sampling and Testing

3.1 QUALITY CONTROL

To assure that materials are in accordance with the Uniform Building Code and the particular project specifications, tests may be required on the mortar, grout, masonry units, and prisms. The following is an excerpt from the 1985 Uniform Building Code Sec. 2405(a) and (b).

U.B.C. Sec. 2405(a), (b)

Quality Control

Sec. 2405. (a) **General.** Quality control shall ensure that materials, construction and workmanship are in compliance with the plans and specifications, and the applicable requirements of this chapter. When required, inspection records shall be maintained and made available to the building official.

(b) **Scope.** Quality control shall include, but is not limited to, assurance that:
1. Masonry units, reinforcement, cement, lime, aggregate and all other materials meet the requirements of the applicable standards of quality and

that they are properly stored and prepared for use.
2. Mortar and grout are properly mixed using specified proportions of ingredients. The method of measuring materials for mortar and grout shall be such that proportions of materials are controlled.
3. Construction details, procedures and workmanship are in accordance with the plans and specifications.
4. Placement, splices and bar diameters are in accordance with the provisions of this chapter and the plans and specifications.

3.2 SAMPLING AND TESTING

Testing should be done in compliance with specifications verified prior to the start of work. Job site tests, if specified, should then be made to confirm the continuing acceptable quality of materials used.

3.2.1 Cone Penetration Test for Consistency of Mortar

The cone penetration test as outlined in ASTM C780 "Standard Method for Preconstruction and Construction Evaluation of Mortars for Plain and Reinforced Unit Masonry" provides a technique for determining the consistency or plasticity of mortar.

Consistency determinations by cone penetration allow controlling additions for all mortars included in the preconstruction test series. Although mortar consistency as measured at the construction site may be a higher penetration value than the preconstruction tests, the cone penetration test serves to standardize water additions for mortar mixes being considered before construction. A cone penetrometer is illustrated in **Figure 3.1**.

Consistency retention by cone penetration using mortar samples provides a means of establishing the early age setting and stiffening characteristics of the mortars.

The cone penetration test method determines the consistency by measuring the penetration of a conical plunger into a mortar sample (see **Figure 3.2**). A cylindrical measure, having an inside diameter of 3'' and a depth of 3-15/32'' ± 1/16'', is filled with mortar in three equal layers. Each layer is spaded 20 times with a metal spatula. The top is levelled and a cone 1-5/8'' in diameter and 3-5/8'' long is released into the mortar. The depth of penetration is measured in millimeters.

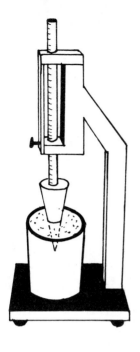

Figure 3.1 Cone penetrometer.

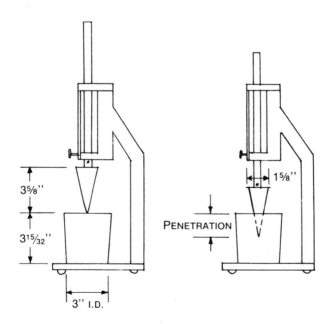

Figure 3.2 Cone penetrometer to test consistency of mortar.

47

Consistency or plasticity of mortar for hollow unit concrete masonry is generally stiffer with a lower cone penetration value than mortar for brick, which generally will be a softer, more plastic mortar. This is because hollow concrete units are heavy and stiff mortar must hold the unit in position without squeezing down. Brick units are light and can be easily moved into position in the plastic mortar.

3.2.2 Field Test for Mortar Strength

It is sometimes necessary to know what the strength is of mortar and grout that is actually used on the project, therefore specimens should be made in the field using job site materials and made in accordance with the test methods of Uniform Building Code Standard No. 24-22 and Section 2405(c)3 of the Uniform Building Code. Refer to **Figure 3.3.**

U.B.C. Sec. 2405(c)3.

3. Mortar Testing. When required by the building official, mortar shall be tested in accordance with U.B.C. Standard No. 24–22.

3.2.3 Field Compressive Test Specimens for Mortar

U.B.C. Standard No. 24–22

Field Compressive Test Specimen for Mortar

Sec. 24.2201. Spread mortar on the masonry units ½ inch to ⅝ inch thick, and allow to stand for one minute, then remove mortar and place in a 2-inch by 4-inch cylinder in two layers, compressing the mortar into the cylinder using a flat end stick or fingers. Lightly tap mold on opposite sides, level off and immediately cover molds and keep them damp until taken to the laboratory. After 48-hours' set, have the laboratory remove molds and place them in the fog room until tested in the damp condition.

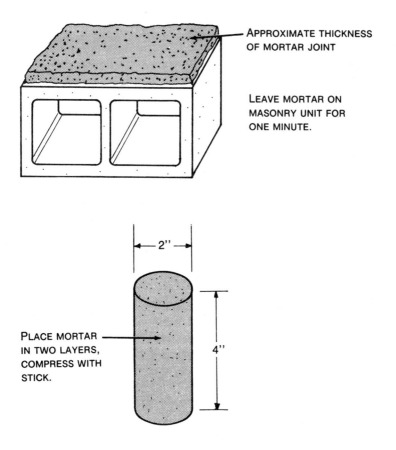

APPROXIMATE THICKNESS
OF MORTAR JOINT

LEAVE MORTAR ON
MASONRY UNIT FOR
ONE MINUTE.

2"

PLACE MORTAR
IN TWO LAYERS,
COMPRESS WITH
STICK.

4"

Figure 3.3 Preparing and making field test mortar specimens.

Proposed Alternate Method: Prepare mortar for specimen by spreading on masonry unit to approximate thickness of mortar joint, place another unit on top, leave there for three minutes, remove top unit and place mortar in mold (see **Figure 3.4**).

The use of this proposed alternate method of preparing the mortar for the test specimen may yield different results compared to the method in U.B.C. Standard Sec. 24.2201. This difference would be due to more water being absorbed by the masonry units as one is placed on top and the mortar is left on for three minutes instead of one minute.

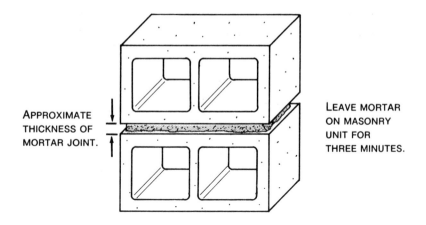

APPROXIMATE THICKNESS OF MORTAR JOINT.

LEAVE MORTAR ON MASONRY UNIT FOR THREE MINUTES.

Figure 3.4 *Alternate method of preparing field mortar for test specimens.*

3.2.4   Mortar Strength Requirements

The Uniform Building Code Standard No. 24-22 states

> **U.B.C. Standard No. 24–22**
>
> **Requirements**
>
> **Section 24.2202.** Each such mortar test specimen shall exhibit a minimum ultimate compressive strength of 1500 pounds per square inch.

It is recommended that each mortar test specimen shall exhibit a minimum 28-day ultimate compressive strength as set forth in **Table 3-A.**

The strength requirement for mortar from U.B.C. Standard No. 24-20 Table No. 24-20-A is based on a 2'' cube, while field test specimens as required by U.B.C. Standard No. 24-22 are based on a 2'' x 4'' cylindrical specimen. The 2'' cube is laboratory prepared

Table 3-A Compressive Strength of Mortar (psi)		
Mortar Type	2" × 4" Cylinder Specimen	2" Cube Specimen
M	2100	2500
S	1500	1800
N	625	750

Lesser periods of time for testing may be used provided the relation between early tested strength and the 28-day strength of the mortar is established.

mortar with a specified flow while the 2" x 4" cylindrical specimen is field mortar.

To obtain an equivalency of a 2" x 4" cylinder field test specimen to a 2" cube specimen, divide the compression test result of the cylinder specimen by 0.83.

If mortar tests are required, the following schedule is suggested.

Take one test per day for three successive work days at the start of the job and store in a moist climate until tested. One test shall consist of three specimens which are made in accordance with Section 24.2201 Field Compression Test Specimens for Mortar, U.B.C. Standard No. 24-22.

After the first three tests, specimens for continuing quality control shall be taken once a week, or for every 2500 square feet of wall, whichever occurs first.

For Type S mortar, compressive strength shall be at least 50% of the 28-day strength after seven days and at least equal to the specified strength after 28 days (Ref. No. 14).

The strength of mortar has some influence on the compressive strength of the masonry prism and wall (Ref. No. 19). It is a convenient control test to assure minimum standards and adequate cementing materials are used in the mortar to provide bond.

3.2.5 Field Tests for Grout

Grout significantly contributes to the strength of the masonry wall and bonds the reinforcing steel into the structural system. Specimens are made in such a way as to duplicate the condition of grout in the wall.

3.2.6 Field Compressive Test Specimens for Grout

Uniform Building Code Standard No. 24-28 outlines the method of making a grout specimen to try to achieve similarity to grout in the wall. The absorptive paper towelling prevents bond of grout to the unit and yet allows the excess moisture to be absorbed into the unit. Refer to **Figure 3.5**.

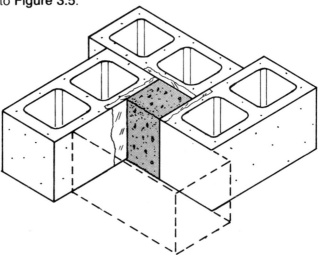

Figure 3.5 *Arrangement of masonry units for making a grout test specimen.*

U.B.C. Standard No. 24–28

Procedure

Sec. 24.2805. (a) Select a level location where the molds can remain undisturbed for 48 hours.

(b) Mold Construction. 1. The mold space should simulate the grout location in the wall. If the grout is placed between two different types of masonry units, both types should be used to construct the mold.

2. Form a square prism space, nominally 3 inches or larger on each side and twice as high as its width, by stacking masonry units of the same type and moisture condition as those being used in the construction. Place wooden blocks, cut to proper size and of the proper thickness or quantity, at the bottom of the space to achieve the necessary height of specimen. Tolerance on space and specimen dimensions shall be within 5 percent of the specimen width. [See **Figure 3.5**.]

3. Line the masonry surfaces that will be in contact with the grout specimen with a permeable material, such as paper towel, to prevent bond to the masonry units.

(c) Measure and record the slump of the grout.

(d) Fill the mold with grout in two layers. Rod each layer 15 times with the tamping rod penetrating ½ inch into the lower layer. Distribute the strokes uniformly over the cross section of the mold.

(e) Level the top surface of the specimen with a straight-edge and cover immediately with a damp absorbent material such as cloth or paper towel. Keep the top surface of the sample damp by wetting the absorbent material and do not disturb the specimen for 48 hours.

(f) Protect the sample from freezing and variations in temperature. Store an indicating maximum-minimum thermometer with the sample and record the maximum and minimum temperatures experienced prior to the time the specimens are placed in the moist room.

(g) Remove the masonry units after 48 hours. Transport field specimens to the laboratory, keeping the specimens damp and in a protective container.

Alternate Methods: Some laboratories and inspectors prepare grout specimens by pouring the grout into the concrete block cells, as shown in **Figure 3.6**. After the grout has set for a few days the

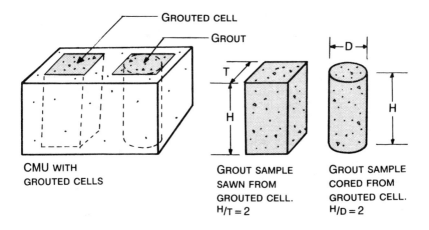

Figure 3.6 *Alternate grout test specimens.*

---

masonry shells and webs are broken off. The grout specimens are either:

a. tested as is and an adjustment made for height and area,

b. sawed into 3-13/16'' x 3-13/16'' x 7-5/8'' prismatic specimens and tested, or

c. a 3'' or 4'' diameter core may be drilled from the grout cell and then tested.

Another method that could be used is to pour the grout into a special concrete block that has three 4'' diameter cells, as shown in **Figure 3.7**. After the grout has set for several days, the block is broken away and three 4'' x 8'' grout specimens are obtained.

The alternate methods of making a grout specimen closely relate to actual field conditions. It is suggested that comparison specimens be made for both the alternate method and standard method, U.B.C. Standard No. 24-28, to establish the relationship between the strength of the grout of specimens made by each method. This relationship, once established for the job, can then be used throughout.

Use of an alternate method of making grout specimens may be subject to question if the test results do not comply with specification requirements if comparison tests are not made.

FIELD SAMPLING OF GROUT USING
ABC GROUT SAMPLE BLOCK

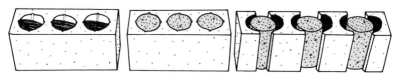

BLOCK IS NOMINAL 6" X 8" X 16" WITH THREE 4" DIAMETER HOLES
MEETS ASTM C90-75 SPECIFICATION.

Figure 3.7 Proprietary grout sample block.

3.2.7   Grout Strength Requirements

The 1988 Uniform Building Code Standard No. 24-29 states, "the grout shall have a minimum compressive strength when tested in accordance with U.B.C. Standard No. 24–28 equal to its specified strength, but not less than 2000 psi."

The minimum compressive strength of 2000 psi is to:

a. insure compatibility with the concrete masonry units

b. provide adequate bond strength of the grout to the reinforcing bars.

This minimum value is satisfactory for masonry construction in which the design strength f'_m = 1500 psi and the masonry unit has a compressive strength of 1900 psi.

U.B.C. Standard No. 24–29

Requirements

Sec. 24.2904.

The grout proportions and any additives shall be based on laboratory or field experience considering the grout ingredients and the masonry units to be used, or the grout shall be proportioned within the limits given in Table No.

24–B of the Uniform Building Code, or the grout shall have a minimum compressive strength when tested in accordance with U.B.C. Standard No. 24–28 equal to its specified strength, but not less than 2000 psi.

It is recommended that the compressive strength of grout in concrete masonry construction be at least equal to 1.33 times the design strength of the masonry assemblage, f'_m. An example of this is that 2000 psi grout is required for a masonry assemblage strength f'_m of 1500 psi.

If grout tests are required, the following schedule is suggested.

At the start of grouting operations, take one test per day for the first three days. The tests shall consist of three specimens which are made in accordance with Section 3.2.5 Field Tests for Grout and U.B.C. Standard No. 24-28.

After the first three tests, specimens for continuing quality control shall be taken once a week or for every 25 cubic yards of grout or for every 2500 square feet of wall, whichever comes first.

For minimum grout strength as required by U.B.C. Standard No. 24-29, the compressive strength shall be at least 1000 psi (Ref. No. 15) after seven days and at least 2000 psi after 28 days. If higher strength grout is required, strength shall be as specified.

3.3 CONCRETE MASONRY UNITS

While most of the tests on concrete masonry units are performed prior to start of work, some random sampling at the job site may be required of concrete masonry units by specifications or by request of the building official, architect or other authorized person. These samples should be truly random, representative, average samples.

The tests that should be conducted are based on the requirements of U.B.C. Standard No. 24-4 Hollow Load-Bearing Concrete Masonry Units and shall meet the requirements for compressive strength, water absorption, and thickness for face shells and webs. In addition, Type I moisture-controlled units shall meet the moisture content requirements based on linear shrinkage of the unit and the humidity conditions at job site or point of use.

The test procedures for compressive strength are outlined in U.B.C. Standard No. 24-7. The test procedures for absorption, weight, moisture content and dimensions are given in ASTM C140-75 (reapproved 1980) Standard Methods of Sampling and Testing Concrete Masonry Units.

See Section 5, Masonry Units, for some of the numerous sizes and types of hollow concrete masonry units.

3.4 PRISM TESTING

3.4.1 General

Prism testing is primarily used when strengths are required higher than the conventional assumed design values allow. Since unusual conditions are frequently involved, it is important that adequate time be allowed for preparing these prisms since retesting could be required. The test is to determine how well different materials work together. The full strength they develop depends on many factors, including workmanship and materials.

The general procedure for making samples, curing and testing is specified in Uniform Building Code standard No. 24-26. The method consists essentially of making sample assemblies of the material to be used in the construction and then testing them to see what capacities that combination of materials will develop. In general, five samples are made and tested prior to starting the work. Then samples are taken at intervals during construction using the same masonry units, mortar, grout and masons used in the construction of the wall.

Care must be exercised in handling the prisms in order to prevent damage before testing. The prisms should be left undisturbed and under moist cover for about two days before being moved to the laboratory. They are then cured moist, as specified, and tested at 28 days. In addition, seven-day tests should be made so the relationship of these strengths to the 28-day strengths can be known. When seven-day tests are made, extrapolation can determine whether results are satisfactorily meeting the 28-day strength requirement.

U.B.C. Standard No. 24-26

Compressive Strength of Masonry Prisms

Sec. 24.2602. Prisms shall be constructed on a flat, level base. Masonry units used in the prism shall be representative of the units used in the corresponding construction. Each prism shall be built in an opened moisture-tight bag which is large enough to enclose and seal the completed prism. The orientation of units, where top and bottom cross sections vary due to taper of the cells, or where the architectural surface of either side of the unit varies, shall be the same orientation as used in the corresponding construction. Prisms shall be a single wythe in thickness and laid up in stack bond.

The length of masonry prisms may be reduced by saw cutting; however, prisms composed of regular shaped hollow units shall have at least one complete cell with one full-width cross web on either end. Prisms composed of irregular-shaped units shall be cut to obtain as symmetrical a cross section as possible. The minimum length of saw-cut prisms shall be 4 inches.

Masonry prisms shall be laid in a full mortar bed (mortar bed both webs and face shells). Mortar shall be representative of that used in the corresponding construction. Mortar joint thickness, the tooling of joints and the method of positioning and aligning units shall be representative of the corresponding construction.

Prisms shall be a minimum of two units in height, but not less than 12 inches. Immediately following the construction of the prism, the moisture-tight bag shall be drawn around the prism and sealed.

Where the corresponding construction is to be solid grouted, prisms shall be solid grouted. Grout shall be representative of that used in the corresponding construction. Grout shall be placed not less than one day nor more than two days following the construction of the prism. Grout consolidation shall be representative of that used in the construction. Additional grout shall be placed in the prism after reconsolidation and settlement due to water loss, but prior to the grout setting. Excess grout shall be screeded off level with the top of the prism. Where open-end units are used, additional masonry units shall be used as forms to confine the grout during placement. Masonry unit forms shall be sufficiently braced to prevent displacement during grouting. Immediately following the grouting operation, the moisture-tight bag shall be drawn around the prism and resealed.

Where the corresponding construction is to be partially grouted, two sets of prisms shall be constructed; one set shall be grouted solid and the other set shall not be grouted.

Where the corresponding construction is of multiwythe composite masonry, masonry prisms representative of each wythe shall be built and tested separately.

Prisms shall be left undisturbed for at least two days after construction.

3.4.2 Standard Prism Tests

Uniform Building Code Standard No. 24-26 Test Method for Compressive Strength of Masonry Prisms is based on the ASTM Standard E447-80. The U.B.C. Standard is given in Section 7 and requires a prism two-units high with one mortar joint, as shown in **Figure 3.8**.

ASTM E447-84 Standard Test Methods for Compressive Strength of Masonry Prisms requires that prisms have at least two mortar joints and can be made up of a half height unit, a full height unit, and a half height unit (see **Figure 3.9**).

3.4.3 Tests of Masonry Prisms

Masonry prisms shall be tested in accordance with U.B.C. Sec. 2405(c)2 and U.B.C. Standard No. 24-26. Refer to **Figures 3.10** and **3.11**.

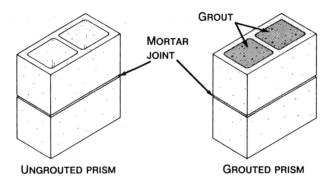

UNGROUTED PRISM   GROUTED PRISM

Figure 3.8 Masonry prism construction for U.B.C. Standard No. 24-26.

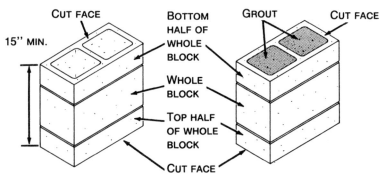

Figure 3.9 Masonry prism construction for ASTM E447-84.

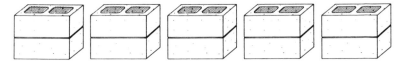

FOR QUALIFYING TEST PRIOR TO CONSTRUCTION, FIVE SPECIMENS ARE RE-
QUIRED FOR ONE TEST.

FOR FIELD CONTROL AS THE PROJECT IS BEING CONSTRUCTED, THREE SPECI-
MENS ARE REQUIRED FOR ONE TEST.

Figure 3.10 Number of specimens for a prism test.

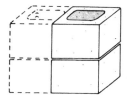

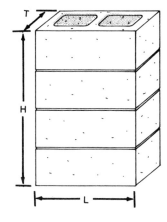

Minimum H = 12"
H/T RATIO MINIMUM 1.5
  MAXIMUM 5.0
L = LENGTH OF UNIT OR PART OF A UNIT
INCLUDING AT LEAST ONE CELL AND
ADJACENT WEB BUT NOT LESS THAN 4".

Figure 3.11 Size of prism specimen.

U.B.C. Sec. 2405(c)2.

2. **Masonry Prism Testing.** When the compressive strength of masonry is specified to be verified by masonry prism tests, the following requirements shall be met:
  A. A set of five masonry prisms shall be built and tested in accordance with U.B.C. Standard No. 24-26 prior to the start of construction.
  B. During construction, a set of three masonry prisms shall be built and tested in accordance with U.B.C. Standard No. 24-26 for each 5,000 square feet of wall area, but not less than one set of three masonry prisms for the project.
  C. The compressive strength of masonry determined in accordance with U.B.C. Standard No. 24-26 for each set of prisms shall equal or exceed f'_m.

3.4.4 Specified Compressive Strength, f'_m

The value of the specified compressive strength (f'_m) shall be based upon either the provisions of Section 2406 (b)1 prism tests or 2406 (b)2, based on field experience, or 2406 (b)3, assumed ultimate compressive strength, from U.B.C. Table No. 24-C.

U.B.C. Sec. 2406(b)

(b) **Specified Compressive Strength of Masonry, f'_m.**
The allowable stresses for the design of masonry shall be based on an f'_m value selected in accordance with one of the following provisions except as modified in Section 2407(h)4:

1. When the compressive strength of masonry is specified in the contract documents to be verified by masonry prism testing in accordance with Section 2405(c)2, the specified compressive strength of masonry, f'_m shall be selected based on an appropriate compressive strength of masonry which can be obtained with the materials and construction practices specified for the project;

U.B.C. Sec. 2406(b)2

2. When there is a masonry prism test record, approved by the building official, or at least 30 masonry prism tests conducted in accordance with U.B.C. Standard No. 24–26 and representative of the corresponding construction, the specified compressive strength of masonry, f'_m may be selected on the basis of this record, but shall not exceed 75 percent of the average value of the prism test record; or

U.B.C. Sec. 2406 (b) 3

3. When neither of the above provisions are met, the specified compressive strength of masonry, f'_m shall be selected from Table No. 24–C.

U.B.C. Sec. 2405 (c) 1

(c) **Quality Control Testing. 1. Sampling and testing of masonry units.**

The sampling and testing of concrete masonry units [Section 2402 (b)5] shall be in accordance with the provisions of U.B.C. Standard No. 24–7.

U.B.C. TABLE NO. 24-C—SPECIFIED COMPRESSIVE STRENGTH OF MASONRY, f'_m (psi)[2] BASED ON SPECIFYING THE COMPRESSIVE STRENGTH OF MASONRY UNITS

Compressive Strength of Clay Masonry Units[1] (psi)	Specified Compressive Strength of Masonry, f'_m	
	Type M or S Mortar[3] (psi)	Type N Mortar[3] (psi)
14,000 or more	5,300	4,400
12,000	4,700	3,800
10,000	4,000	3,300
8,000	3,350	2,700
6,000	2,700	2,200
4,000	2,000	1,600
Compressive Strength of Concrete Masonry Units[4] (psi)	Specified Compressive Strength of Masonry, f'_m	
	Type M or S Mortar[3] (psi)	Type N Mortar[3] (psi)
4,800 or more	3,000	2,800
3,750	2,500	2,350
2,800	2,000	1,850
1,900	1,500	1,350
1,250	1,000	950

[1] Compressive strength of solid clay masonry units is based on gross area. Compressive strength of hollow clay masonry units is based on minimum net area. Values may be interpolated.

[2] Assumed assemblage. The specified compressive strength of masonry f'_m is based on gross area strength when using solid units or solid grouted masonry and net area strength when using ungrouted hollow units.

[3] Mortar for unit masonry, proportion specification, as specified in Table No. 24-A. These values apply to portland cement-lime mortars without added air-entraining materials.

[4] Net area compressive strength of concrete masonry units is determined in accordance with U.B.C. Standard No. 24-7. Values may be interpolated. In grouted concrete masonry the compressive strength of grout shall be equal to or greater than the compressive strength of the concrete masonry units.

3.5 SUMMARY

Specifications and verification requirements for the State of California, Title 24, the City of Los Angeles and the Uniform Building Code are summarized in **Table 3-B**. **Table 3-C** is a summary of masonry inspection requirements, and **Table 3-D** summarizes mortar mix design requirements. [Ref. 16]

Table 3B Specification and Verification Requirements[1] for Masonry and Masonry Materials *

Material Category	Material Type & Property	Specification Requirements UBC & ACI Requirements	Additional Verification Requirements[2]		
			CAC T24 Part 2	(Rev.) 1985 COLA Bldg. Code	1988 UBC
Building, Facing, Hollow Brick	Burned Clay Brick	UBC Standard No. 24-1 (ASTM C62, C216, C652)	Test Prior to Use Grade NW Not Permitted	Not Required	Not Required
Sand-Lime Brick	Sand-Lime Brick	UBC Standard No. 24-2 (ASTM C73)	Material Not Permitted	Not Required	Not Required
Clay Load Bearing Tile	Burned, Hollow Clay Tile	UBC Standard No. 24-8 (ASTM C34 & C112)	Material Not Permitted	Not Required	Not Required
Structural Clay, Non Load Brg. Tile	Burned Hollow Clay Tile	UBC Standard No. 24-9 (ASTM C56)	Material Not Permitted	Not Required	Not Required

Cast Stone	Portland Concrete Trim & Facing	UBC Standard No. 24-13 (ACI 704)	Test Prior to Use	Not Required	Not Required
Unburned Clay Units	Adobe Block	UBC Standard No. 24-14	Material Not Permitted	Not Required	Not Required
Wire	Steel Wire	UBC Standard No. 24-15 (ASTM A82)	Test Prior to Use	Test Prior to Use	Not Required
In-cement	Reinforcing Bars	UBC Standard No. 26-4 (ASTM A615, A616, A617, A706, A767 & A775)	Test Prior to Use[3]	Test Prior to Use[4]	Not Required
Quicklime	Unslaked Lime	UBC Standard No. 24-17 (ASTM C5)	Test as Required	Not Required	Not Required
Hydrated Lime	Slaked Lime	UBC Standard No. 24-18 (ASTM C207)	Test as Required	Not Required	Not Required

*Ref. No. 17

65

Table 3B Specification and Verification Requirements[1] for Masonry and Masonry Materials

Material Category	Material Type & Property	Specification Requirements UBC & ACI Requirements	Additional Verification Requirements[2]		
			CAC T24 Part 2	(Rev.) 1985 COLA Bldg. Code	1988 UBC
Grout	Portland Cement Grout	UBC Section 2403(d) COLA Section 91.2403(r), T-24, Part 2, Section 2403(s) & UBC Standards No. 24-28 & 24-29 (ASTM C1019, C476)	Proportioned per Code Field Test for Compressive Strength Required	Proportioned per Code[5]	Proportioned per Code[5]
Masonry Cement	Cement (Any Binder)	UBC Standard No. 24-16 (ASTM C91)	Material Not Permitted	Material Not Permitted	Material Not Permitted in Seismic Zones 2, 3, or 4
Cores	Sample of Masonry	T-24 Section 2405A(c)C.	Compression & Shear Tests per Section 2405A(c)C[6]	Not Required Except by Chapter 88	Not Required

Admixtures		UBC Building Code Section 2403(e), COLA Section 91.2403(q), T-24, Section 2-2403(r)	See Footnote 7	Per Research Approval for Specific Admixture[8]	See Footnotes 7 and 9
Portland Cement	Portland Cement	UBC Standard No. 26-1 (ASTM C150)	Test as Required	Not Required	Not Required
Metal Ties & Anchors	Steel Connectors	UBC Building Code Section 2402(b)7	Not Required	Not Required	Not Required
Water	Water	UBC Building Code Section 2402(b)11	Not Required	Not Required	Not Required
Aggregate for Grout	Aggregates for Grout	UBC Standard No. 24-23 (ASTM C404)	Test Prior to Use	Test Prior to Use	Not Required
Aggregate for Mortar	Aggregates for Mortar	UBC Standard No. 24-21 (ASTM C144)	Test Prior to Use	Test Prior to Use	Not Required

Table 3B *Specification and Verification Requirements[1] for Masonry and Masonry Materials*

Material Category	Material Type & Property	Specification Requirements UBC & ACI Requirements	Additional Verification Requirements[2]		
			CAC T24 Part 2	(Rev.) 1985 COLA Bldg. Code	1988 UBC
Mortar	Proportion Spec. or Property Spec.	UBC Standard No. 24-20 (ASTM C161 & C270)	Proportioned per Code Field Test for Compressive Strength Required[10]	By Proportioning per Code or Properties Verified by Complete Tests Prior to Use[5]	By Proportioning per Code or Properties Verified by Complete Tests Prior to Use[5]
Mortar	Ready Mixed	Proposed ASTM Standard, ASTM C157, ASTM E514, ASTM C666, UBC Standard No. 24-20, ASTM C952, ASTM C876	Subject to Prior Approval of Agency	Subject to Prior Approval of City	Subject to Prior Approval of Building Official
Mortar	Field Test	UBC Standard No. 24-22	Per Section 2404(d)[10]	As Required by City	As Required by Building Official

Material	Type	Standard			
Concrete Masonry Units	Hollow Load Bearing	UBC Standard No. 24-4 (ASTM C90)	Grade S Not Permitted Complete Test Prior to Use	Not Required	Not Required
Concrete Masonry Units	Drying Shrinkage	UBC Standard No. 24-27 (ASTM C426)	Not Required	Not Required	Not Required
Concrete Masonry Units	Non Load Bearing	UBC Standard No. 24-6 (ASTM C129)	Complete Test Prior to Use	Not Required	Not Required
Masonry Prisms	Qualification Testing	UBC Standard No. 24-26 (ASTM E447)	Prism Tests if Spec. f'm is Greater than 1500 psi with Approved Quality Control Methods	Preliminary Tests reqd. to Determine f'm Unless Assumed Values per Sec. 91.2404(c)3 are Used — Field Tests per Every 5000 sq. ft. of Wall. 3 Tests min.	Preliminary Tests reqd. to Determine f'm Unless Assumed Values per Table 24-D are Used or Field Experience per Sec. 2406(b) 3 is Submitted. — Field Tests per Every 5000 sq. ft. of Wall, 1 min.

Table 3B Specification and Verification Requirements[1] for Masonry and Masonry Materials

Material Category	Material Type & Property	Specification Requirements UBC & ACI Requirements	Additional Verification Requirements[2]		
			CAC T24 Part 2	(Rev.) 1985 COLA Bldg. Code	1988 UBC
Cement	Plastic	U.B.C. Standard No. 26-1, Part 1 Except Insoluble Residue, Air Entrainment, & Additions Subsequent to Calcination	Special Approval is Necessary Sec. T24-2, Sec. 2-2403(p)	Plasticizing Agents Require Approval Prior to Use	Plasticizing Agents Require Approval Prior to Use
Concrete Brick	Concrete Building Brick	UBC Standard No. 24-3 (ASTM C55)	Complete Test Prior To Use	Not Required	Not Required
Ceramic Structural Facing Tile	Load Bearing Tile	UBC Standard No. 24-25 (ASTM C126)	Complete Test Prior to Use	Not Required	Not Required
Brick	Sampling and Testing	UBC Standard No. 24-24 (ASTM C67)	Not Required	Not Required	Not Required

Gypsum Concrete	Gypsum Concrete	1985 UBC Standard No. 26-15; 1982 UBC Standard No. 24-12	Complete Test Prior to Use	Compressive Test Required Other Tests as Required by City or Specifications	Compressive Test Required Other Tests as Required by Jurisdiction or Specifications

[1] The specifications listed fall into two classifications. The U.B.C. Standards are requirements generally mandated by the building code enforcement agency and are essentially ASTM Standards and ACI Standards modified to conform to such requirements. The ASTM Standards and the ACI Standards are industry standards used to inform the buyer of the conditions under which materials are furnished.

[2] Additional verification requirements are specific tests and/or analyses required by the building code enforcement agency in addition to the U.B.C. specification requirements. These tests and/or analyses generally are based on the U.B.C. specification indicated, and generally are conducted by independent agents such as testing laboratories.

[3] Tensile and bend tests per each 10 tons or fraction thereof of each size of reinforcement in each lot. Where manufacturer's name, or heat identification number, or manufacturer's chemical analysis is unknown, testing frequency must be increased to each 2-1/2 tons or fraction thereof.

[4] Tensile and bend tests per each 25 tons or fraction thereof of each size of reinforcement in each lot. Where manufacturer's name, or heat identification number, or manufacturer's chemical analysis is unkrown, testing frequency must be increased to each 5 tons or fraction thereof.

[5] Field tests for compressive strength may be required, by jurisdiction or specifications.

[6] Two cores minimum, with at least one core taken from every 5000 square feet of floor area, or fraction thereof.

[7] Admixtures require prior approval by the building official before addition to grout and mortar.

[8] For color, only pure mineral oxides are permitted.

[9] For color, only pure mineral oxides, carbon black, or synthetic colors are permitted. Carbon black is limited to 3 percent of cement weight.

[10] Test sampling consists of at least one sample taken on each of three consecutive working days at beginning of masonry work, and at least at weekly intervals thereafter.

Table 3-C *Summary of Inspection Requirements**

		MASONRY INSPECTION REQUIREMENT	
C.A.C Title 24 Part 2 Sec. 2-2416A		1985 City of Los Angeles Code Sections 91.0309, 91.0310, 91.2405, 91.2406 and 91.2407	1988 U.B.C., Section 306 and 2405
1. Continuous inspection during laying and grouting of masonry including placing of reinforcement and taking test samples. Check all materials, details of construction and construction procedures. 2. Continuous inspection during placing of reinforced gypsum concrete including		1. City inspector to inspect grouted masonry when vertical reinforcing steel is in place and other reinforcing steel distributed and ready for placing but before any units are laid up. 2. City inspector to inspect grouted masonry after the masonry is laid up and prior to grouting when continuous inspection by a registered deputy building inspector is not required. 3. Continuous inspection by a registered deputy building inspector is required for all masonry construction when higher stresses are used in design. (All masonry designed at higher than one-half stress.) 4. Continuous inspection by a registered	1. Called inspections of masonry construction may be required by the building official. 2. Continuous inspection by a special inspector is required during the preparation of masonry wall prisms, sampling and placing (laying) of all masonry units, placing of reinforcement, inspection of grout space immediately prior to closing cleanouts and all grouting operations. This special inspection includes assurance that all materials meet the applicable quality standards, proportioning and mixing of mortar and grout, construction details, procedures and workmanship, and the placement of reinforcement including splices. (Special inspection need not be

72

placing of reinforcement and taking test samples. Check all materials, details of construction and construction procedures.	deputy building inspector is required for the grouting of all high-lift grouted construction regardless of design stresses.	provided when design stresses have been adjusted to permit noncontinuous inspection—one-half stress or less (2406(c)1.) with maximum limits (2407(h)4.). Some inspections may be made on a periodic basis and satisfy the requirements of continuous inspection, provided this periodic scheduled inspection is performed as outlined in the project plans and specifications and approved by the building official. *NOTE:* The plans shall describe the required strengths of masonry materials and inspection requirements for which all parts of the structure were designed.

*Ref. No. 16

Table 3-D *Mortar Mix Design Requirements**

1985 CAC T24, Part 2	1985 L.A. City Building Code	1988 U.B.C.[1]
1. Proportions and strength shall conform to Section 2-2403, T-24-2	1. In accordance with the materials and proportions in Table No. 24-A Compressive Strength must comply with U.B.C. Standard No. 24-20. 2. In accordance with property specifications in U.B.C. Standard No. 24-20.	1. Proportions established by laboratory or field experience with the mortar ingredients and masonry units to be used. Compressive strength must comply with U.B.C. Standard No. 24-20. 2. In accordance with the materials and proportions in Table 24-A. Compressive strength must comply with U.B.C. Standard No. 24-20. 3. In accordance with property specifications in U.B.C. Standard No. 24-20.

[1] The mixing of mortar grout in a mechanical mixer shall be for a period of 3-10 minutes. Mortar may be re-tempered.

*Ref. No. 16

For quality construction, mortar must be mixed to the proper consistency, as shown above.

Construction

4.1 GENERAL

Inspection is most important during actual construction. The inspector's job is to assure that all work performed is done according to the approved plans and specifications, and the materials are as specified and used correctly.

4.2 PREPARATION OF FOUNDATION AND SITE

Prior to laying the first course of concrete masonry, the concrete surfaces shall be clean and free of laitance, loose aggregate, grease, or anything that will prevent the mortar from bonding properly. It shall be rough to provide good bond between foundation concrete and mortar and grout.

U.B.C. Sec. 2407 (h) 4. H.

H. Mortar joints between masonry and concrete. Concrete abutting structural masonry such as at starter courses or at wall intersections not designed as true separation joints shall be roughened to a full amplitude of ⅛ inch and shall be bonded to the masonry per the requirements of this chapter as if it were masonry. Unless keys or proper reinforcement are provided, vertical joints

---

> as per Section 2407(b)2 shall be considered to be stack bond and the reinforcement as required for stack bond shall extend through the joint and be anchored into the concrete.

Surfaces shall be level and at a correct grade so that the initial bed joint shall not be less than 1/4'' nor more than 1'' in height.

> **U.B.C. Sec. 2407 (d). 1**
>
> (d) **Support of Masonry.** 1. **Vertical support.** Structural members providing vertical support of masonry shall provide a bearing surface on which the initial bed joint shall not be less than ¼ inch nor more than 1 inch and shall be of noncumbustible material, except where masonry is a nonstructural decorative feature or wearing surface.

Reinforcing steel dowels shall be properly placed. The specified size and length shall be checked and if they need to be bent, they may be bent at a slope of no more than 1'' horizontally per six inches of height, as stated in U.B.C. Sec. 2607 (i) 1. A.

If any of the site conditions or layout are improper, masonry work should not begin until cleared by the governing authority.

The first course on the foundation should have all webs and face shells set in mortar for full bearing. The mortar, however, must not project more than 1/2 inch nor more than the vertical height of the mortar joint into the cells that are to contain grout, as shown in **Figure 4.1**. The grout must have direct contact and bearing on the foundation or slab.

4.3 MATERIALS, HANDLING, STORAGE AND PREPARATION

Given below are Uniform Building Code requirements, Sec. 2404(a) General and Sec. 2404(b) Materials-Handling, Storage and Preparation. **Figures 4.2** through **4.7** illustrate these requirements.

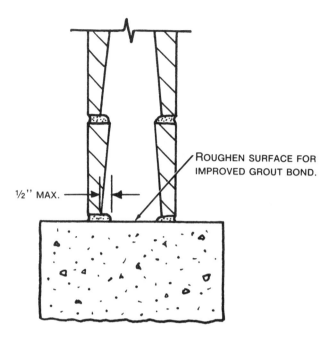

ROUGHEN SURFACE FOR
IMPROVED GROUT BOND.

½" MAX.

Figure 4.1 First course mortar joint.

Figure 4.2 Proper delivery and storage of masonry units.

Figure 4.3 *Reinforcing steel.*

Figure 4.4 *Concrete masonry units should not be wetted.*

Figure 4.5 Properly stored cement sacks.

Figure 4.6 Using measuring box for accurate proportioning.

Figure 4.7 *Mixing mortar.*

U.B.C. Sec. 2404 (a) and (b)

Construction

Sec. **2404.** (a) **General.** Masonry shall be constructed according to the provisions of this section.

(b) **Materials—Handling, Storage and Preparation.** All materials shall comply with applicable requirements of Section 2402. Storage, handling and preparation at the site shall conform also to the following:

1. Masonry materials shall be stored so that at the time of use the materials are clean and structurally suitable for the intended use.

2. All metal reinforcement shall be free from loose rust and other coatings that would inhibit reinforcing bond.

3. At the time of laying, burned clay units and sand lime units shall have a rate of absorption not exceeding .025 ounce per square inch during a period of one minute. In the absorption test the surface of the unit shall be held ⅛ inch below the surface of the water.

4. Concrete masonry units shall not be wetted unless otherwise approved.

5. Materials shall be stored in a manner such that deterioration or intrusion of foreign materials is prevented and that the material will be capable of meeting applicable requirements at the time of mixing.

6. The method of measuring materials for mortar and grout shall be such that proportions of the materials can be controlled.

7. Mortar or grout mixed at the jobsite shall be mixed for a period of time not less 3 minutes nor more than 10 minutes in a mechanical mixer with the amount of water required to provide the desired workability. Hand mixing of small amounts of mortar is permitted. Mortar may be retempered. Mortar or grout which has hardened or stiffened due to hydration of the cement shall not be used, but under no case shall mortar be used two and one-half hours, nor grout used one and one-half hours, after the initial mixing water has been added to the dry ingredients at the jobsite.

4.4 PLACEMENT AND LAYOUT

4.4.1 General

All dimensions, locations of all wall openings, positions of vertical reinforcing, methods of grouting, mortar mixes, patterns of bond, and the general sequence of operations should be decided prior to laying the first course of masonry.

Where no bond pattern is shown, the wall should be laid in straight uniform courses with alternate vertical joints aligning (called running bond or common bond, shown in **Figure 4.8**). Proper alignment of the vertical cells gives maximum size openings for pouring

Figure 4.8 *Running or common bond masonry.*

grout in vertically reinforced cells and reduces ledges or projections that may impede the flow of grout.

If units are layed in stack bond, shown in **Figure 4.9**, give particular attention to proper type and placement of reinforcing steel or metal ties and joint reinforcement used to provide the mechanical bond.

Unless specified otherwise, vertical and horizontal mortar joints for precision units shall be 3/8'' ± 1/8'', and for slumped units shall be 1/2'' ± 1/4''.

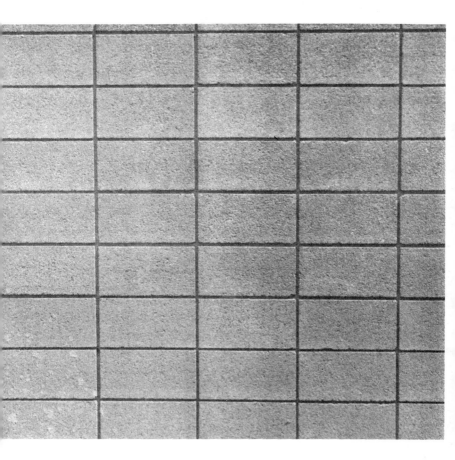

Figure 4.9 Masonry laid up in stack bond.

4.4.2 Placing Masonry Units

U.B.C. Sec. 2404 (d)

(d) **Placing Masonry Units. 1. Mortar.** The mortar shall be sufficiently plastic and units shall be placed with sufficient pressure to extrude mortar from the joint and produce a tight joint. Deep furrowing which produces voids shall not be used.

The initial bed joint thickness shall be not less than ¼ inch nor more than 1 inch; subsequent bed joints shall be not less than ¼ inch nor more than ⅝ inch in thickness.

2. Surfaces. Surfaces to be in contact with mortar or grout shall be clean and free of deleterious materials.

3. Solid masonry units. Solid masonry units shall have full head and bed joints. [See Figure 4.10.]

U.B.C. Sec. 2404 (d) 4.

4. Hollow masonry units. All head and bed joints shall be filled solidly with mortar for a distance in from the face of the unit not less than the thickness of the shell. [See Figure 4.11.]

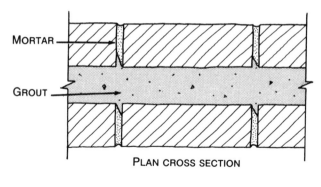

PLAN CROSS SECTION

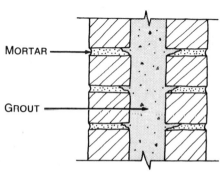

Figure 4.10
Grout flows into head joint and bed joint for full joints.

ELEVATION CROSS SECTION

U.B.C. Sec. 2404 (d) 4.

Head joints of open-end units with beveled ends need not be mortared. The beveled ends shall form a grout key which permits grout within ⅝ inch of the face of the unit. The units shall be tightly butted to prevent leakage of grout. [See **Figure 4.12.**]

Pilasters (in the wall columns) are laid up at the same time as the wall, taking care to place the pilaster ties as required.

No unit should be moved after setting as this breaks the mortar bond. Should moving of a unit be necessary, the mortar should be removed and the unit set in fresh mortar.

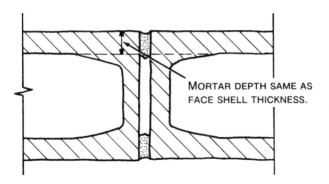

MORTAR DEPTH SAME AS FACE SHELL THICKNESS.

Figure 4.11 *Hollow masonry unit head joints.*

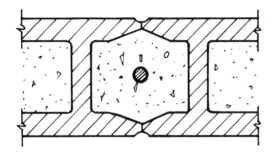

Figure 4.12 *Speed block mortarless head joints.*

4.4.3 Typical Layout of CMU Walls

A. Corner Details are illustrated in **Figures** 4.13 and 4.14.

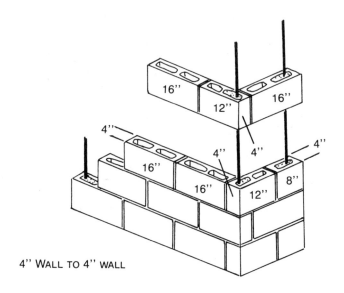

4" WALL TO 4" WALL

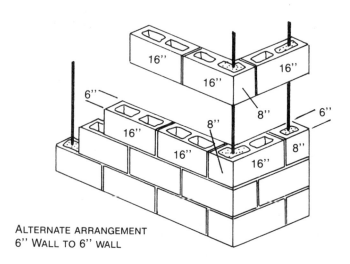

ALTERNATE ARRANGEMENT
6" WALL TO 6" WALL

Figure 4.13 *Arrangement of masonry units for corners.*

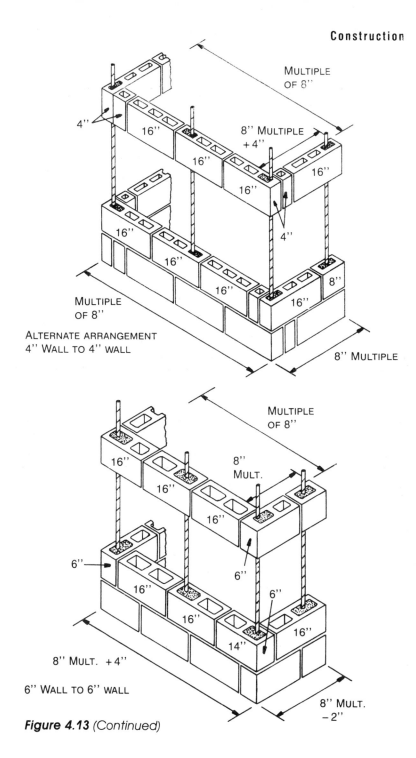

Figure 4.13 (Continued)

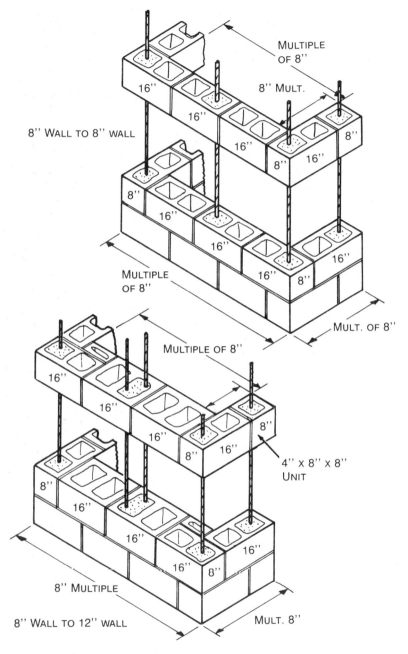

Figure 4.14 *Arrangement of masonry units for corners.*

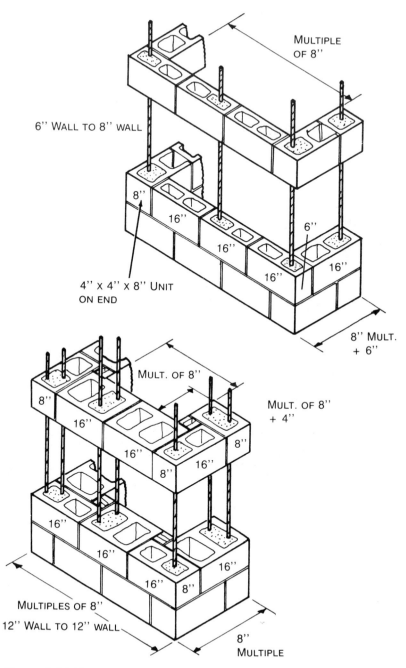

MULTIPLE OF 8''

6'' WALL TO 8'' WALL

8''

16''

16''

16''

6''

16''

4'' x 4'' x 8'' UNIT ON END

8'' MULT. + 6''

MULT. OF 8''

MULT. OF 8'' + 4''

8''

16''

16''

8''

16''

8''

16''

16''

16''

16''

8''

MULTIPLES OF 8''

12'' WALL TO 12'' WALL

8'' MULTIPLE

Figure 4.14 (Continued)

B. Typical Layout of Pilasters is shown in
Figures 4.15 through 4.17.

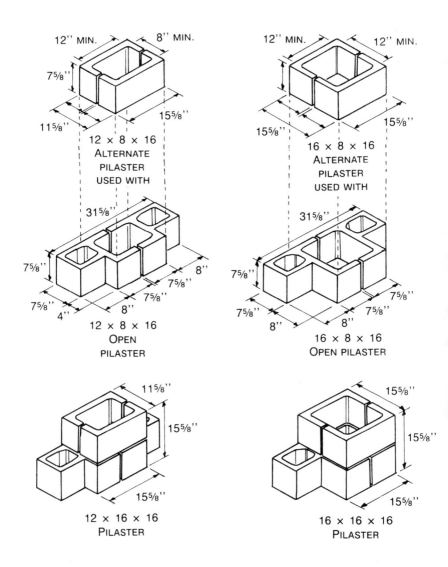

Figure 4.15 *Arrangement of units for pilaster.*

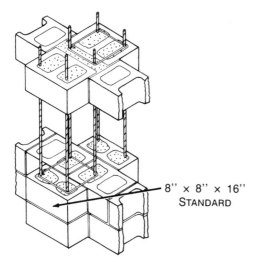

8'' × 8'' × 16''
STANDARD

FOUR NO. 6 & TWO NO 5
BARS MINIMUM
¼'' TIE AT 16'' CENTERS
28'' × 16'' STANDARD
IN 12'' WALL

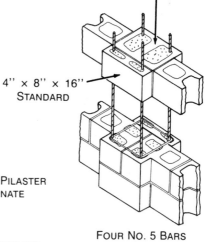

12'' × 8'' × 16'' STD.

4'' × 8'' × 16''
STANDARD

FOUR NO. 5 BARS
MINIMUM
RECOMMENDED MAXIMUM
FOUR NO. 10 BARS
¼'' TIE AT 16'' CENTERS
16'' × 16'' CENTERED

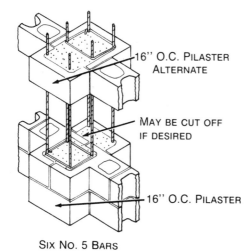

16'' O.C. PILASTER
ALTERNATE

MAY BE CUT OFF
IF DESIRED

16'' O.C. PILASTER

SIX NO. 5 BARS
MINIMUM
RECOMMENDED MAXIMUM
SIX NO. 11 BARS
¼'' TIE AT 16'' CENTERS
24'' × 16'' OPEN
CENTER-CENTERED

Figure 4.16 Pilaster details.

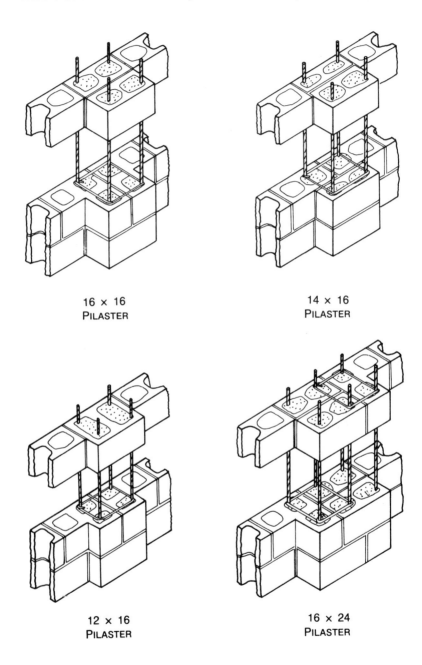

16 × 16
PILASTER

14 × 16
PILASTER

12 × 16
PILASTER

16 × 24
PILASTER

Figure 4.17 Pilaster details.

C. Typical Connections of Intersecting Walls and Embedded Columns are shown in Figures 4.18 through 4.20.

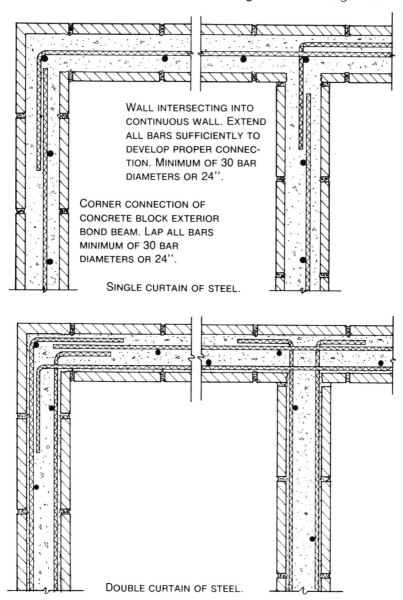

WALL INTERSECTING INTO CONTINUOUS WALL. EXTEND ALL BARS SUFFICIENTLY TO DEVELOP PROPER CONNECTION. MINIMUM OF 30 BAR DIAMETERS OR 24".

CORNER CONNECTION OF CONCRETE BLOCK EXTERIOR BOND BEAM. LAP ALL BARS MINIMUM OF 30 BAR DIAMETERS OR 24".

SINGLE CURTAIN OF STEEL.

DOUBLE CURTAIN OF STEEL.

Figure 4.18 Typical intersecting wall connections.

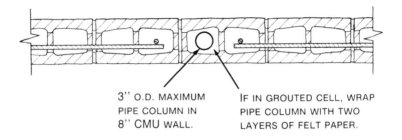

3" O.D. MAXIMUM
PIPE COLUMN IN
8" CMU WALL.

IF IN GROUTED CELL, WRAP
PIPE COLUMN WITH TWO
LAYERS OF FELT PAPER.

Figure 4.19 *Embedded steel columns in masonry wall.*

D. Lintel and Bond Beam.

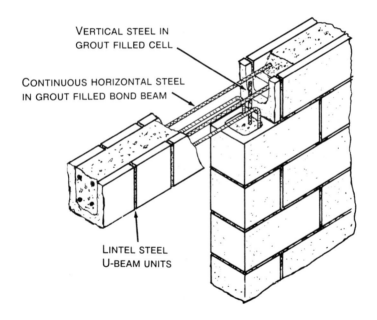

VERTICAL STEEL IN
GROUT FILLED CELL

CONTINUOUS HORIZONTAL STEEL
IN GROUT FILLED BOND BEAM

LINTEL STEEL
U-BEAM UNITS

Figure 4.20 *Lintel and bond beam detail.*

E. **Arrangement of Open End Units** is shown in
Figures 4.21 and 4.22.

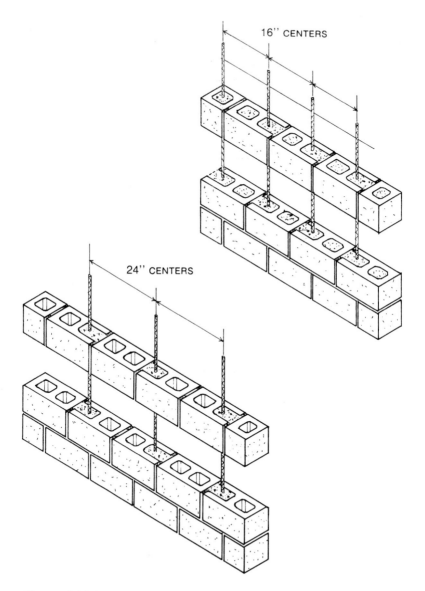

16" CENTERS

24" CENTERS

Figure 4.21 Arrangement of steel and
open end units—16" and 24" spacing.

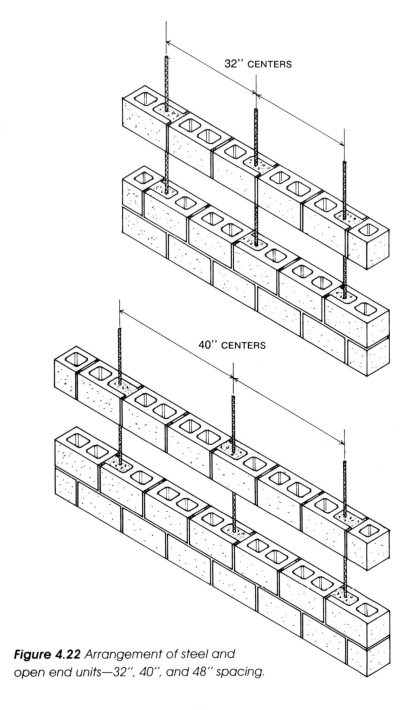

Figure 4.22 *Arrangement of steel and open end units—32", 40", and 48" spacing.*

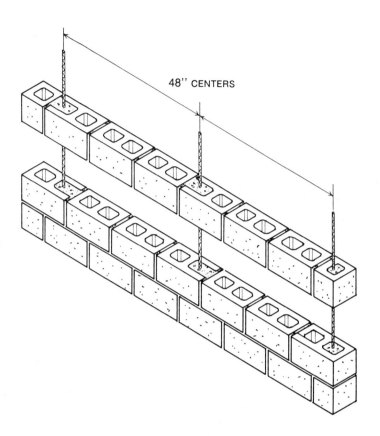

48'' CENTERS

Figure 4.22 (Continued)

F. Typical Wall Assembly is shown in Figure 4.23.

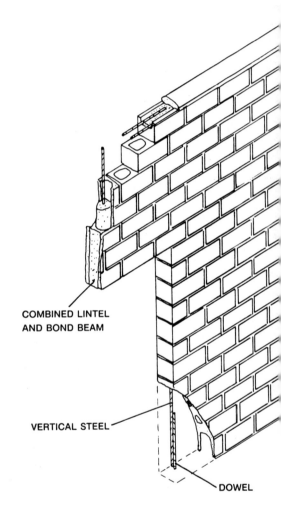

COMBINED LINTEL
AND BOND BEAM

VERTICAL STEEL

DOWEL

Figure 4.23 Wall assembly.

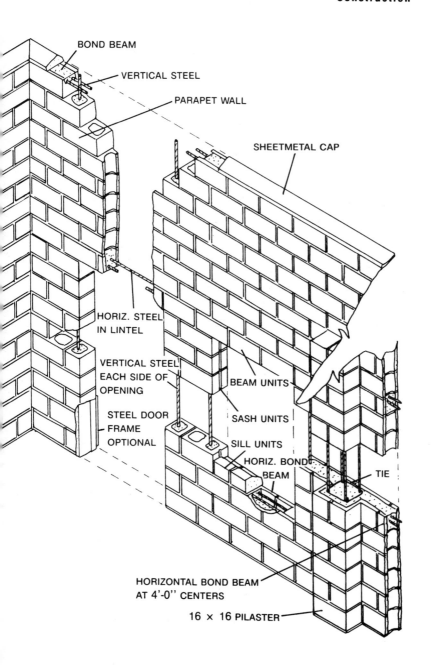

BOND BEAM

VERTICAL STEEL

PARAPET WALL

SHEETMETAL CAP

HORIZ. STEEL IN LINTEL

VERTICAL STEEL EACH SIDE OF OPENING

STEEL DOOR FRAME OPTIONAL

BEAM UNITS

SASH UNITS

SILL UNITS

HORIZ. BOND BEAM

TIE

HORIZONTAL BOND BEAM AT 4'-0'' CENTERS

16 × 16 PILASTER

101

4.5 MORTAR JOINTS

Mortar is used in the joints between masonry units. The horizontal joint is the bed joint and the vertical joint is the head joint. Mortar is the bedding material that allows the units to be placed level, plumb and in proper position. Mortar is also the sealing material between masonry units. The exposed surface of the mortar is finished in a number of ways, as illustrated in **Figures 4.24** and **4.25**.

Concave, V joints and weathered joints are recommended for exterior masonry. Tooling the joints requires pressure, which compresses the mortar, creating a tight bond between the mortar and the unit and provides a dense surface for weatherproofing and seals the interface between mortar and masonry unit.

Mortar joints for interiors may be the same as exterior joints or they may be raked, extruded or weeping, struck or flush cut. These joints increase the chance for water leakage since the small ledges allow water to collect and migrate into the wall at the mortar unit interface. These joints require special attention and tooling to improve water tightness and are not recommended for exterior work.

Flush cut joints may be used where the finished surface is to be plastered, painted or covered with wallpaper.

Special effect joints that are sometimes used are beaded and grapevine. These are for simulating old style masonry.

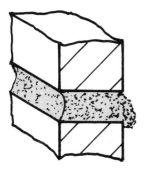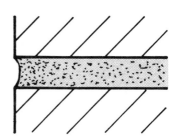

CONCAVE JOINT
MOST COMMON JOINT USED, TOOLING WORKS THE MORTAR TIGHT INTO THE JOINT TO PRODUCE A GOOD WEATHER JOINT. PATTERN IS EMPHASIZED AND SMALL IRREGULARITIES IN LAYING ARE CONCEALED.

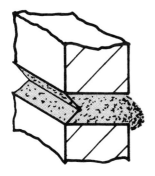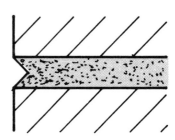

"V" JOINT
TOOLING WORKS THE MORTAR TIGHT AND PROVIDES A GOOD WEATHER JOINT. USED TO EMPHASIZE JOINTS AND CONCEAL SMALL IRREGULARITIES IN LAYING AND PROVIDE A LINE IN CENTER OF MORTAR JOINT.

Figure 4.24 Types of recommended mortar joints, providing the best weather protection.

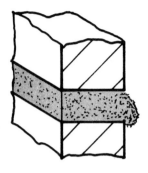 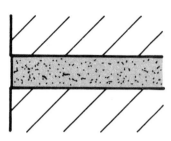

FLUSH JOINT

USE WHERE WALL IS TO BE PLASTERED OR WHERE IT IS DESIRED TO HIDE JOINTS UNDER PAINT. SPECIAL CARE IS REQUIRED TO MAKE JOINT WEATHER-PROOF. MORTAR JOINTS MUST BE COMPRESSED TO ASSURE INTIMATE CONTACT WITH THE BLOCK.

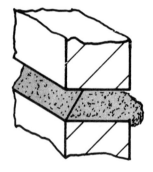 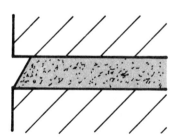

WEATHERED JOINT

USE TO EMPHASIZE HORIZONTAL JOINTS. ACCEPTABLE WEATHER JOINT WITH PROPER TOOLING.

Figure 4.24 *(Continued) Types of acceptable mortar joints (weather joint possible with proper tooling).*

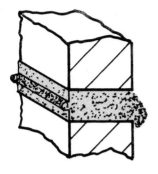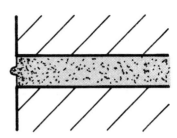

BEADED JOINT
SPECIAL EFFECT, POOR EXTERIOR WEATHER JOINT BECAUSE OF EXPOSED
LEDGE—NOT RECOMMENDED.

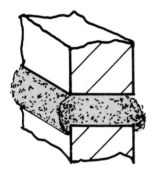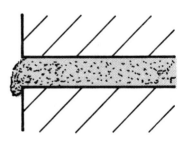

SQUEEZED JOINT
PROVIDES A RUSTIC, HIGH TEXTURE LOOK. SATISFACTORY INDOORS AND EX-
TERIOR FENCES. NOT RECOMMENDED FOR EXTERIOR BUILDING WALLS.

Figure 4.25 Types of non-weather mortar joints, for special
effects only.

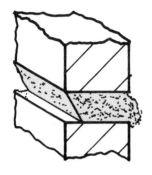
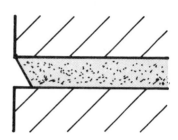

STRUCK JOINT

USE TO EMPHASIZE HORIZONTAL JOINTS. POOR WEATHER JOINT—NOT RECOMMENDED AS WATER WILL PENETRATE ON LOWER EDGE.

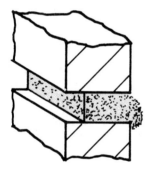
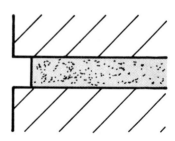

RAKED JOINT

STRONGLY EMPHASIZES JOINTS. POOR WEATHER JOINT—NOT RECOMMENDED UNLESS TOOLED AT BOTTOM OF MORTAR JOINT IF EXPOSED TO WEATHER.

Figure 4.25 (Continued) Types of non-weather mortar joints.

4.6 REINFORCING STEEL

Reinforcing steel is the material that imparts ductility, added strength and toughness to masonry structures. It is one of the primary components for lateral force-resistant design and construction.

4.6.1 Maximum Size and Amount of Reinforcing Steel

U.B.C. Sec. 2409 (e)

(e) **Reinforcing Requirements and Details. 1. Maximum reinforcing size.** The maximum size of reinforcing shall be No. 11. Maximum steel area in cells shall be 6 percent of the cell area without splices and 12 percent of the cell area with splices. [See Table 4-A.]

The suggested maximum size and amount is to reduce congestion and facilitate grouting of the cells. Splices increase the congestion and therefore it is suggested that they be staggered.

4.6.2 Spacing of Steel in Walls

Placing the steel reinforcement in the proper locations is critical. For a structure to resist wind and seismic loads, the steel reinforcement must be where it can function properly. Uniform Building Code Sec. 2409 (e) 2 specifically states certain reinforcement locations.

U.B.C. Sec. 2409 (e) 2

2. **Spacing of longitudinal reinforcement.** The clear distance between parallel bars, except in columns, shall be not less than the nominal diameter of the bars nor 1 inch, except that bars in a splice may be in contact. This clear distance requirement applies to the clear distance between a contact splice and adjacent splices or bars. [See Figures 4.26 and 4.27.]

Table 4-A Maximum Size of Reinforcing Bars in CMU Cells * (Walls)

Nominal Thickness of CMU	Approx. Cell Size	Approx. Cell Area	Max. Steel Area—6% of Cell Area	Bar Selection	Suggested Maximum	
					Steel Area—4% of Cell Area	Bar Selection
4"	1¾" × 5½"	10 sq. in.	0.6 sq. in.	1–#7	0.4 sq. in.	1–No. 6
6"	3¼" × 5½"	18 sq. in.	1.1 sq. in.	1–#9	0.7 sq. in.	1–No. 8
8"	5" × 5¼"	26 sq. in.	1.6 sq. in.	1#11	1.0 sq. in.	1–No. 9
10"	7" × 5¼"	37 sq. in.	2.2 sq. in.	2#9	1.5 sq. in.	1 No. 11 or 2 No. 7
12"	9" × 5¼"	47 sq. in.	2.8 sq. in.	2#10	1.9 sq. in.	2 No. 9

* Area of steel in cells may be doubled for lapped spliced bars.

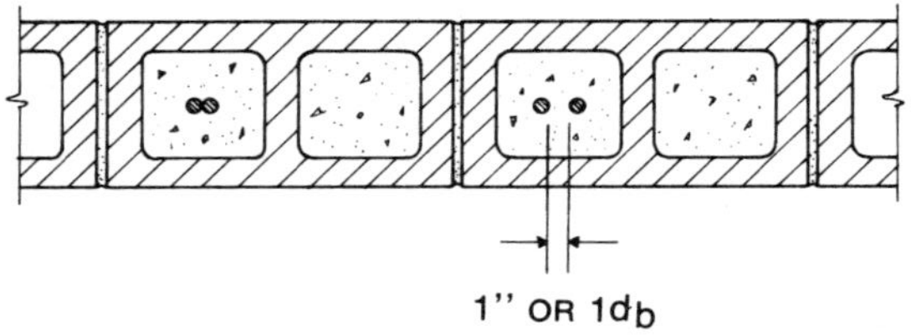

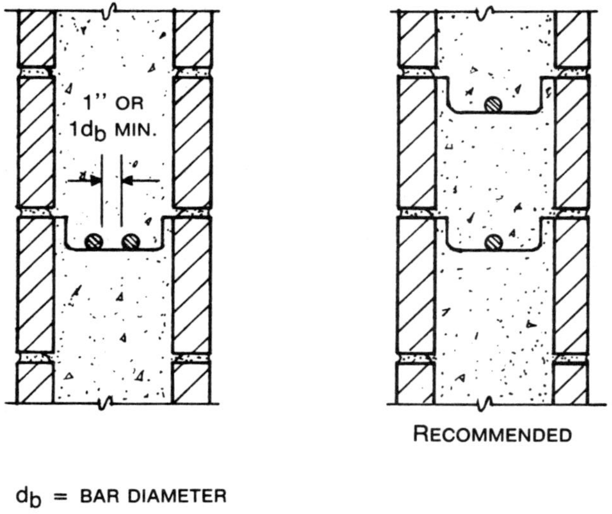

Figure 4.27 Spacing of horizontal reinforcing in masonry wall.

d_b = BAR DIAMETER

RECOMMENDED

1" OR $1d_b$ MIN.

Figure 4.26 Spacing of vertical reinforcing in cell.

1" OR $1d_b$

4.6.3 Clearances of Steel and Masonry

For a reinforced masonry wall to function properly it is important to have the reinforcing steel completely surrounded by grout. This requires that the bars be kept a minimum distance from the masonry to allow the grout to flow around the steel and bond the concrete block and steel together. U.B.C. Sec. 2409 (e) 2 states the minimum requirements.

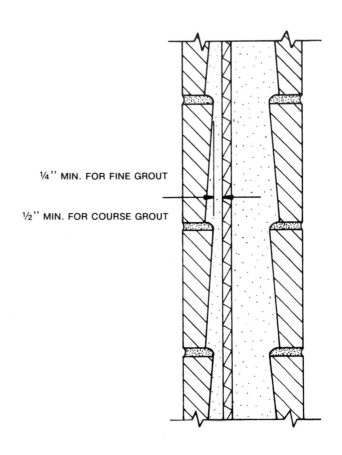

¼" MIN. FOR FINE GROUT

½" MIN. FOR COURSE GROUT

Figure 4.28 *Clearance of reinforcing steel.*

U.B.C. Sec. 2409 (e) 2.

The clear distance between the surface of a bar and any surface of a masonry unit shall be not less than ¼ inch for fine grout and ½ inch for coarse grout. Cross webs of hollow units may be used as support for horizontal reinforcement. [See **Figures** 4.28 and 4.29.]

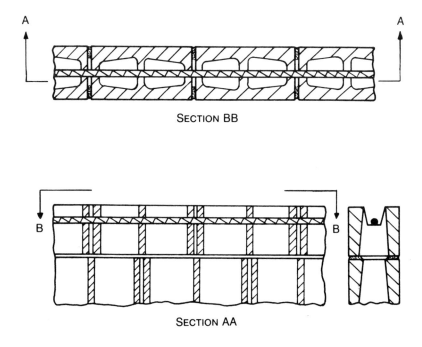

SECTION BB

SECTION AA

Figure 4.29 *Support of reinforcing steel.*

U.B.C. Sec. 2409 (e) 2.

All reinforcing bars, except joint reinforcing, shall be completely embedded in mortar or grout and have a minimum cover, including the masonry unit, of at least ¾ inch, 1½ inches of cover when exposed to weather and 2 inches of cover when exposed to soil. [See Figure 4.30.]

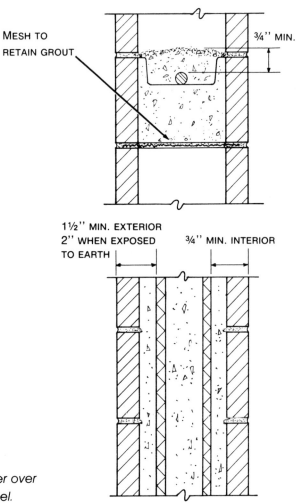

MESH TO RETAIN GROUT

¾'' MIN.

1½'' MIN. EXTERIOR
2'' WHEN EXPOSED
TO EARTH

¾'' MIN. INTERIOR

Figure 4.30
Minimum cover over reinforcing steel.

Wiring the vertical steel to dowels projecting from the foundation is satisfactory if the dowels are in the prescribed location. If they are not, the dowels can be bent to properly position them, as shown in **Figure** 4.31. However, the vertical steel can lap the dowels without the bars being wired together. In fact, they can be separated by several inches and transmit force between them.

If necessary, due to improper location or failure to install dowels, new dowels may have to be put in. These can be installed

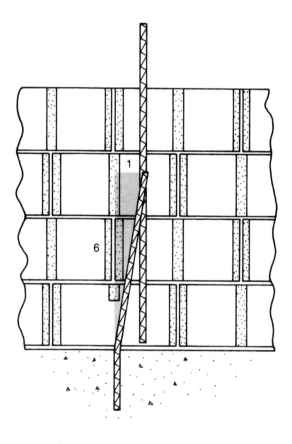

Figure 4.31
Slope for bending reinforcing steel into position.

by several methods, such as drilling and dry packing mortar around a reinforcing bar, drilling and installing red heads, using cinch anchors, anchor shields, etc.

Vertical steel may be held in place by reinforcing bar positioners. These wire positioners will locate the bar in the proper position, e.g., center, to one side, one bar on each side, etc, and will also secure it from moving during grouting of the wall (see **Figure 4.32**).

4.6.4 Securing Reinforcing Steel

U.B.C. Sec. 2404 (e) states the requirements for securing and locating tolerances for reinforcing steel.

U.B.C. Sec. 2404 (e)

(e) Reinforcement Placing. Reinforcing details shall conform to the requirements of Section 2409 (e). Metal reinforcement shall be located in accordance with the plans and specifications. Reinforcement shall be secured against displacement prior to grouting by wire positioners or other suitable devices at intervals not exceeding 200 bar diameters nor 10 feet. [See **Table 4-B**.]

Table 4-B *Maximum Intervals for Securing Reinforcing Bars*

Bar Size	Secured Intervals	Bar Size	Secured Intervals
#3	6'3''	# 8	10'0''
4	8'4''	9	10'0''
5	10'0''	10	10'0''
6	10'0''	11	10'0''
7	10'0''	12	10'0''

As mentioned in section 4.2 reinforcing dowels may be bent at a slope of no more than one inch horizontally per six inches of height, as stated in U.B.C. Sec. 2607 (i) 1. A.

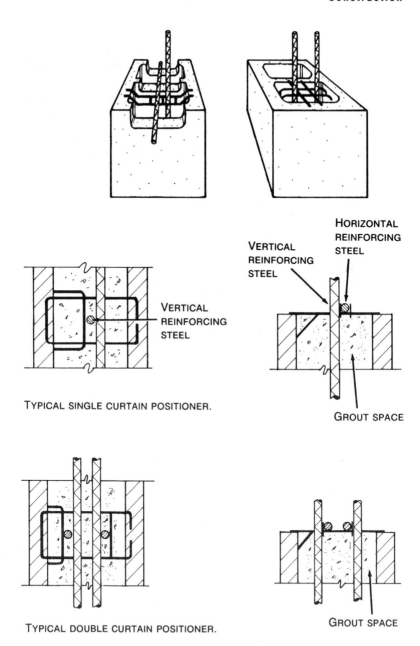

TYPICAL SINGLE CURTAIN POSITIONER.

VERTICAL
REINFORCING
STEEL

HORIZONTAL
REINFORCING
STEEL

VERTICAL
REINFORCING
STEEL

GROUT SPACE

TYPICAL DOUBLE CURTAIN POSITIONER.

GROUT SPACE

Figure 4.32 Bar positioners are also used to locate and hold vertical and horizontal steel.

> **U.B.C. Sec. 2607 (i) 1. A.**
>
> **(i) Special Reinforcing Details for Columns. 1. Offset bars.** Offset bent longitudinal bars shall conform to the following:
>
> A. Slope of inclined portion of an offset bar with axis of column shall not exceed 1 in 6.

4.6.5 Location Tolerances of Bars

Proper location of structural reinforcing steel is important for safe and adequate performance. To assure proper location, U.B.C. Sec. 2404 (e) provides tolerances permitted for placement of bars.

> **U.B.C. Sec. 2404 (e)**
>
> Tolerances for the placement of steel in walls and flexural elements shall be plus or minus ½ inch for d equal to 8 inches or less, plus or minus one inch for d equal to 24 inches or less but greater than 8 inches, and plus or minus 1¼ inch for d greater than 24 inches.
>
> Tolerance for longitudinal location of reinforcement shall be plus or minus 2 inches. [See Table 4-C and Figure 4.33.]

4.6.6 Lap Splices, Reinforcing Bars

Reinforcing bars and joint reinforcing are typically delivered to construction job sites in uniform lengths which can be easily handled by one man. When reinforcing bars meet in a wall they must be connected in some fashion so that all of the stresses can be transferred from one bar to the other. This is usually accomplished by lapping or splicing the bars.

Table 4-C Tolerances for Placing Reinforcement	
Distance, d, from face of CMU to the center of reinforcing	Allowable Tolerance
d ≤ 8''	± ½''
8'' < d ≤ 24''	± 1''
d > 24''	± 1¼''

TOLERANCE
SEE TABLE 4-C

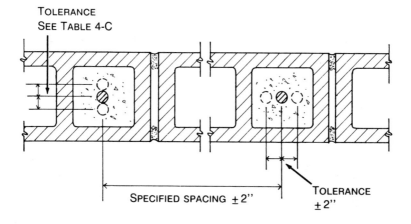

SPECIFIED SPACING ± 2''

TOLERANCE
± 2''

Figure 4.33 Typical tolerances for placement of reinforcing bars in a cell for an 8'' CMU.

Splices may be made only at such points and in such manner that the structural strength of the member will not be reduced. Lapped splices shall provide sufficient lap to transfer the working stress of the reinforcement by bond and shear.

Bars that are spliced shall be in the same cell as the bar to be spliced and not in an adjacent cell.

U.B.C. Sec. 2409 (e) 6.

6. Splices. The amount of lap of lapped splices shall be sufficient to transfer the allowable stress of the reinforcement as in Section 2409(e)3. In no case shall the length of the lapped splice be less than 30 bar diameters for compression and 40 bar diameters for tension.

Welded or mechanical connections shall develop 125 percent of the specified yield strength of the bar in tension.

EXCEPTION: For compression bars in columns that are not part of the seismic system and are not subject to flexure, the compressive strength only need be developed.

When adjacent splices in grouted masonry are from 0 to 3 inches apart, the lap length shall be increased by 1.3 times.

The last sentence in U.B.C. Sec. 2409 (e) 6 is for multiple splices in a cell. Thus, if two bars are both spliced in one cell and are less than 3 inches apart, the splice length shall be 30% longer. See **Figure 4.34**.

The 1985 Uniform Building Code specifies required lap lengths (called development length—l_d) for steel reinforcement.

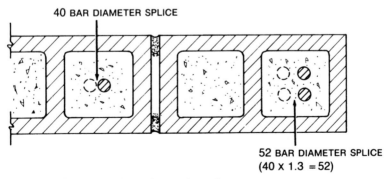

40 BAR DIAMETER SPLICE

52 BAR DIAMETER SPLICE
(40 x 1.3 = 52)

Figure 4.34 *Lap splice of steel in cell.*

3. Anchorage of flexural reinforcement. A. The tension or compression in any bar at any section must be developed on each side of that section by the required development length. The development length of the bar may be achieved by a combination of an embedment length, anchorage or, for tension only, hooks.

The required development length for deformed bars or deformed wire shall be calculated by:

$$l_d = 0.002 \, d_b f_s \text{ for bars in tension} \quad \ldots \ldots \quad (9\text{--}11)$$
$$l_d = 0.0015 \, d_b f_s \text{ for bars in compression} \ldots \quad (9\text{--}12)$$

Development length for smooth bars shall be 2.0 times the length by Formula (9–11).

The development length for deformed bars shall be as follows with a minimum length of 12 inches.

For Grade 40 reinforcing bars in
   tension . . . .$l_d = 40 \, d_b$

For Grade 40 reinforcing bars in
   compression . . . .$l_d = 30 \, d_b$

For Grade 60 reinforcing bars in
   tension . . . .$l_d = 48 \, d_b$

For Grade 60 reinforcing bars in
   compression . . . .$l_d = 36 \, d_b$

     where d_b = diameter of the reinforcing bar.

Refer to **Table 4-D**, Length of Lap (inches).

4.6.7 Lap Splices, Joint Reinforcing

The Uniform Building Code has specified lap lengths and splices for joint reinforcement.

U.B.C. Sec. 2404 (h)

(h) **Joint Steel.** Wire joint reinforcement used in the design as principal reinforcing in hollow unit construction shall be continuous between supports unless splices are made by lapping:

1. Fifty-four wire diameters in a grouted cell, or

2. Seventy-five wire diameters in the mortared bed joint, or

3. In alternate bed joints of running bond masonry a distance not less than 54 diameters plus twice the spacing of the bed joints, or

4. As required by calculation and specific location in areas of minimum stress, such as points of inflection.

Side wires shall be deformed and shall conform to U.B.C. Standard No. 24-15, Part I, Joint Reinforcement for Masonry. [See Table 4-E.]

4.6.8   Coverage and Layout of Joint Reinforcing Steel

A. Coverage. All longitudinal wires shall be completely embedded in mortar. Joint reinforcement embedded in horizontal mortar joints shall have not less than 5/8″ mortar coverage from the exposed face. Refer to **Figure 4.35**.

U.B.C. Sec. 2407 (f)

(f) **Protection of Ties, Bolts and Joint Reinforcement.** A minimum of ⅝-inch mortar cover shall be provided between ties or joint reinforcement and any exposed face. The thickness of grout or mortar between masonry units and joint reinforcement or bolts shall be not less than ¼ inch, except that ¼ inch or smaller diameter reinforcement or bolts may be placed in bed joints which are at least twice the thickness of the reinforcement.

Table 4-D   Length of Lap (inches)

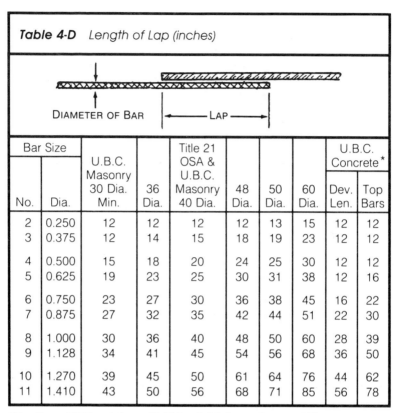

DIAMETER OF BAR |←——— LAP ———→|

Bar Size		U.B.C. Masonry 30 Dia. Min.	36 Dia.	Title 21 OSA & U.B.C. Masonry 40 Dia.	48 Dia.	50 Dia.	60 Dia.	U.B.C. Concrete *	
No.	Dia.							Dev. Len.	Top Bars
2	0.250	12	12	12	12	13	15	12	12
3	0.375	12	14	15	18	19	23	12	12
4	0.500	15	18	20	24	25	30	12	12
5	0.625	19	23	25	30	31	38	12	16
6	0.750	23	27	30	36	38	45	16	22
7	0.875	27	32	35	42	44	51	22	30
8	1.000	30	36	40	48	50	60	28	39
9	1.128	34	41	45	54	56	68	36	50
10	1.270	39	45	50	61	64	76	44	62
11	1.410	43	50	56	68	71	85	56	78

*For f'_c = 2000 psi; f_y = 40,000 psi; Minimum Lap = 12"

Table 4-E   Length of Lap of Wire Joint Reinforcing (inches)

Wire Size		1	2	U.B.C. 3	
No.	Dia.	54 Dia.	75 Dia.	Vertical Spacing 8"	Vertical Spacing 16"
9	.1483	8	12	16	24
8	.1620	9	13	17	25
3/16"	.1875	11	14	19	27

1 Grouted Cell;
2 Mortared Joint;
3 Alternate bed joints.

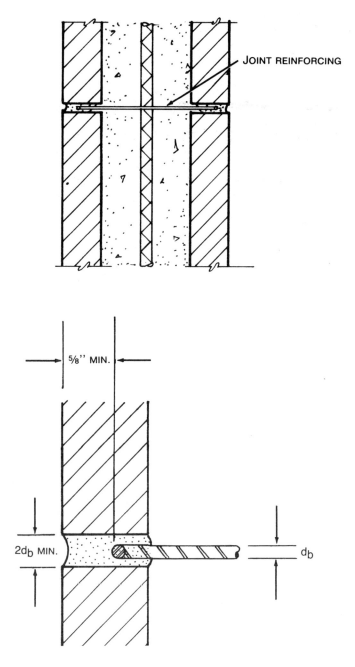

Figure 4.35 Cover over joint reinforcing.

B. Layout of Intersecting Walls. Figure 4.36 illustrates a plan for joint reinforcing showing intersecting walls and alternative lapping. **Figure 4.37** shows typical joint reinforcing.

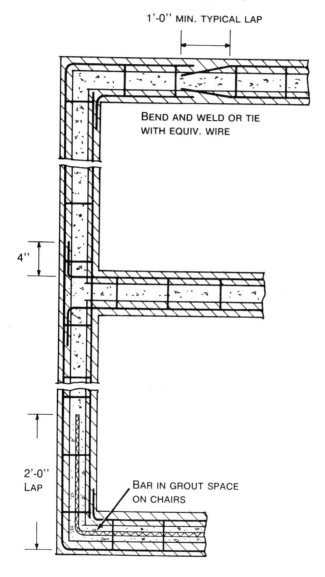

1'-0'' MIN. TYPICAL LAP

BEND AND WELD OR TIE
WITH EQUIV. WIRE

4''

2'-0''
LAP

BAR IN GROUT SPACE
ON CHAIRS

Figure 4.36 *Plan of joint reinforcing showing intersection and alternate lapping.*

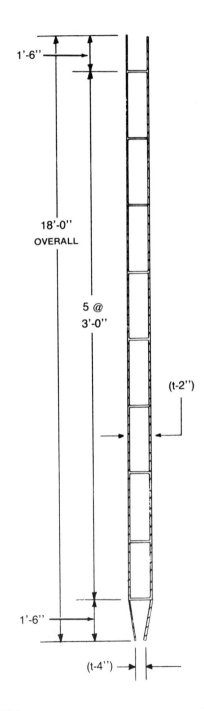

Figure 4.37
Typical joint reinforcing.

4.6.9 Hooks and Bends in Reinforcing Bars

The general requirements for hooks and bends in reinforcing bars are stated in U.B.C. Sec. 2409 (e) 5.

U.B.C. Sec. 2409 (e) 5

5. **Hooks.** A term "standard hook" shall mean one of the following:

(i) A 180–degree turn plus extension of at least 4 bar diameters but not less than 2½ inches at free end of bar.

(ii) A 90–degree turn plus extension of at least 12 bar diameters at free end of bar.

(iii) For stirrup and tie anchorage only, either a 90–degree or a 135–degree turn, plus an extension of at least 6 bar diameters but not less than 2½ inches at the free end of the bar.

B. The diameter of bend measured on the inside of the bar, other than for stirrups and ties, shall be not less than values specified in Table No. 24–F. Except for Grade 40 bars in sizes No. 3 through No. 11, inclusive, the minimum diameter of bend shall be not less than 5 bar diameters.

C. Inside diameter of bend for No. 4 or smaller stirrups and ties shall be not less than 4 bar diameters. Inside diameter of bend for No. 5 or larger stirrups and ties shall be not less than given in Table No. 24–F.

U.B.C. TABLE NO. 24–F—MINIMUM DIAMETERS OF BEND

Bar Size	Minimum Diameter
No. 3 through No. 8	6 bar diameters
No. 9 through No. 11	8 bar diameters

Table 4-F   Minimum Diameters of Bend (Grade 40 Steel)

Bar Size	Minimum Diameter
No. 3 through No. 11	5 bar diameters

Table 4-G   Standard Hook and Bend

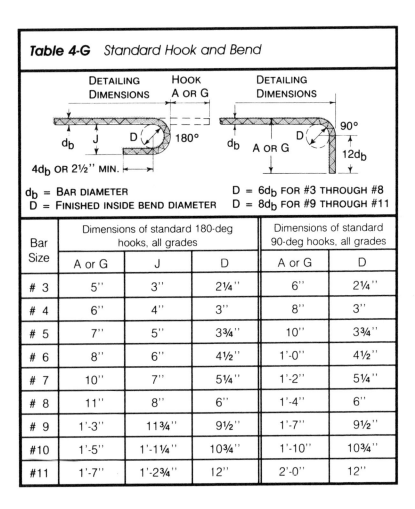

d_b = BAR DIAMETER
D = FINISHED INSIDE BEND DIAMETER
D = 6d_b FOR #3 THROUGH #8
D = 8d_b FOR #9 THROUGH #11

Bar Size	Dimensions of standard 180-deg hooks, all grades			Dimensions of standard 90-deg hooks, all grades	
	A or G	J	D	A or G	D
# 3	5"	3"	2¼"	6"	2¼"
# 4	6"	4"	3"	8"	3"
# 5	7"	5"	3¾"	10"	3¾"
# 6	8"	6"	4½"	1'-0"	4½"
# 7	10"	7"	5¼"	1'-2"	5¼"
# 8	11"	8"	6"	1'-4"	6"
# 9	1'-3"	11¾"	9½"	1'-7"	9½"
#10	1'-5"	1'-1¼"	10¾"	1'-10"	10¾"
#11	1'-7"	1'-2¾"	12"	2'-0"	12"

U.B.C. Table No. 24-F refers to grade 60 steel. **Table 4-F** refers to grade 40 steel. Standard hooks and bends are shown in **Table 4-G**.

4.6.10 Anchorage of Shear Reinforcing Steel

Reinforcing steel to resist shear loads must be anchored according to U.B.C. Sec. 2409 (e) 4. Refer to **Figure 4.38**.

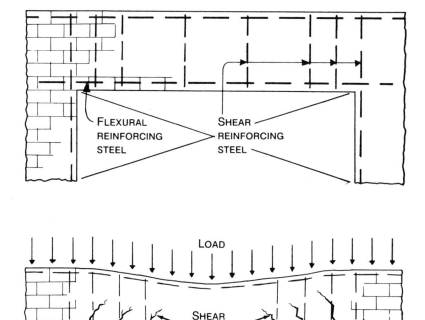

Figure 4.38 *Example of shear reinforcing in beams, and possible crack patterns from excessive loads.*

U.B.C. Sec. 2409 (e) 4.

4. **Anchorage of shear reinforcement.** A. Single separate bars used as shear reinforcement shall be anchored at each end by one of the following methods:

(i) Hooking tightly around the longitudinal reinforcement through 180 degrees.

(ii) Embedment above or below the mid-depth of the beam on the compression side a distance sufficient to develop the stress in the bar for plain or deformed bars.

(iii) By a standard hook [see Section 2409(e)5] considered as developing 7500 psi, plus embedment sufficient to develop the remainder of the stress to which the bar is subjected. The effective embedded length shall not be assumed to exceed the distance between the mid-depth of the beam and the tangent of the hook.

B. The ends of bars forming a single U or multiple U stirrup shall be anchored by one of the methods of Section 2409(e)4A or shall be bent through an angle of at least 90 degrees tightly around a longitudinal reinforcing bar not less in diameter than the stirrup bar, and shall project beyond the bend at least 12 diameters of the stirrup.

C. The loops or closed ends of simple U or multiple U stirrups shall be anchored by bending around the longitudinal reinforcement through an angle of at least 90 degrees and project beyond the end of the bend at least 12 diameters of the stirrup bar. [See **Figures 4.39** and 4.40.]

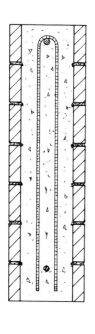

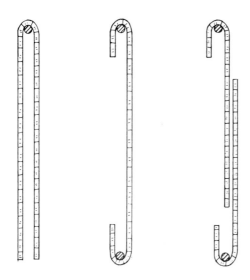

Figure 4.39 Shear Steel for beams.

129

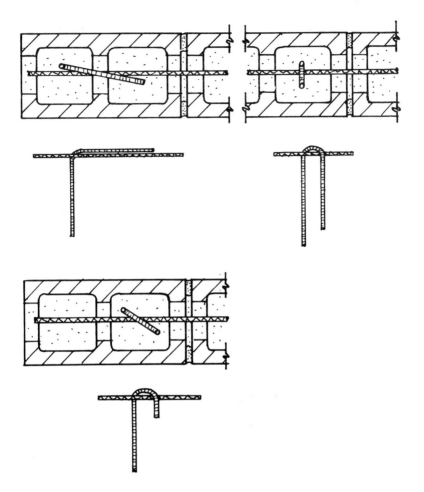

Figure 4.40 Details of beam shear reinforcing.

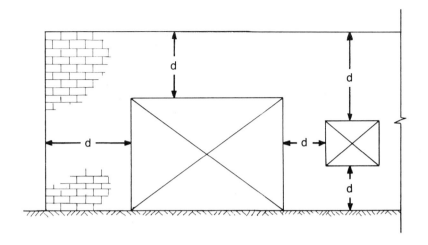

Figure 4.41 Elements, d, for spacing of shear reinforcing. Maximum spacing 48″.

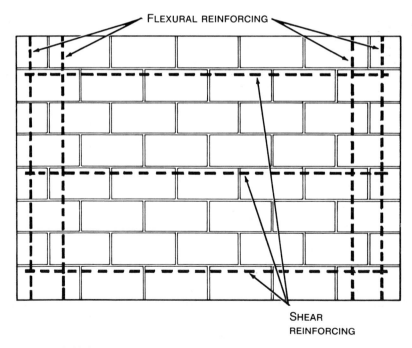

Figure 4.42 Reinforcing for shear walls.

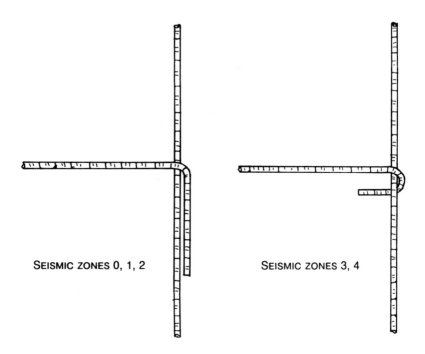

Figure 4.43 *Details of reinforcing for shear walls.*

4.7   COLUMN REINFORCEMENT

4.7.1   Vertical Reinforcement

Steel reinforcement for concrete masonry must conform to the same clearances and tolerances as other masonry, with some additional requirements.

> **U.B.C. Sec. 2409 (b) 5. A.**
>
> **5. Reinforcement for columns. A. Vertical reinforcement.** The area of vertical reinforcement shall be not less than .005 A_e and not more than 0.04 A_e. At least four No. 3 bars shall be provided. [See Figure 4.44.]

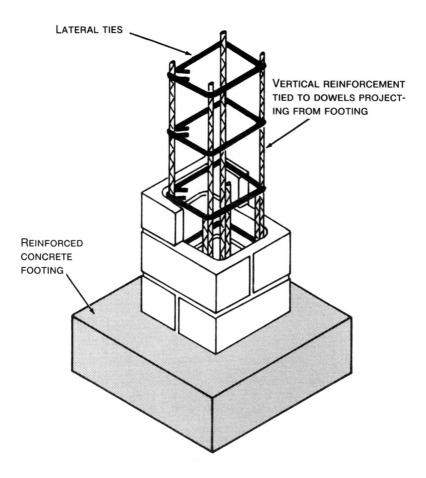

LATERAL TIES

VERTICAL REINFORCEMENT
TIED TO DOWELS PROJECT-
ING FROM FOOTING

REINFORCED
CONCRETE
FOOTING

Figure 4.44 Construction of reinforced concrete masonry column.

U.B.C. Sec. 2409 (e) 2.

The minimum clear distance between parallel bars in columns shall be two and one half times the bar diameter.

4.7.2 Reinforcing Tie Details

A. Lateral Tie Details are shown in **Figures** 4.45 and 4.46. Also refer to **Tables 4-H** and **4-I**.

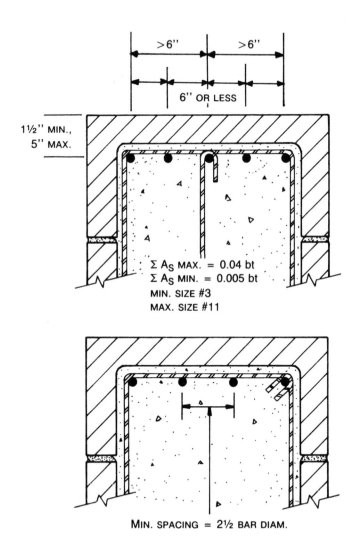

Figure 4.45 *Tie details.*

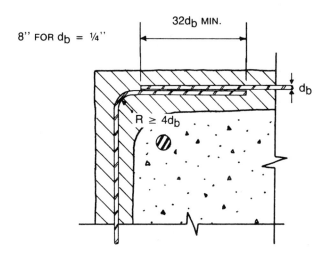

Figure 4.46 *Tie anchorage in bed joint, seismic zones 3 and 4.*

Table 4-H Tie Spacing—16 Bar Diameters	
Compression Steel Bar No.	Maximum Tie Spacing
3	6''
4	8''
5	10''
6	12''
7	14''
8	16''
9	18''
10	18''
11	18''

Note: Maximum tie spacing, 16 bar diameters or 18'' or least dimension of column.

Table 4-I Tie Spacing—48 Tie Diameters	
Tie Steel Bar No.	Maximum Tie Spacing
¼'' 3	12'' 18''
4 5	18'' 18''

Note: #2 (¼'') ties at 8'' spacing is equivalent to #3 (⅜'') tie at 16'' spacing. Maximum tie spacing is 48 tie diameters or 18'' or least dimension of column.

U.B.C. Sec. 2409 (b) 5.B

B. Lateral ties. All longitudinal bars for columns shall be enclosed by lateral ties. Lateral support shall be provided to the longitudinal bars by the corner of a complete tie having an included angle of not more than 135 degrees or by a hook at the end of a tie. The corner bars shall have such support provided by a complete tie enclosing the longitudinal bars. Alternate longitudinal bars shall have such lateral support provided by ties and no bar shall be farther than 6 inches from such laterally supported bar.

Lateral ties and longitudinal bars shall be placed not less than 1½ inches and not more than 5 inches from the surface of the column. Lateral ties may be against the longitudinal bars or placed in the horizontal bed joints if the requirements of Section 2407 (g) are met. Spacing of ties shall be not less than 16 longitudinal bar diameters, 48 tie diameters or the least dimension of the column but not more than 18 inches.

Ties shall be at least ¼ inch in diameter for No. 7 or smaller longitudinal bars and No. 3 for larger longitudinal bars. Ties less than ⅜ inch in diameter may be used for longitudinal bars larger than No. 7, provided the total cross-sectional area of such smaller ties crossing a longitudinal plane is equal to that of the larger ties at their required spacing.

U.B.C. Sec. 2407 (h) 4. G. Hooks

EXCEPTION: Where the bars are placed in the horizontal bed joints, the hook shall consist of a 90–degree bend having a radius of not less than four bar diameters plus an extension of 32 bar diameters.

U.B.C. Sec. 2407 (h) 4. C.

C. **Column reinforcement.** The spacing of column ties shall not be more than: 8 inches the full height for columns stressed by tensile or compressive axial overturning forces due to the seismic loads of Section 2312; 8 inches for the tops and bottoms of all other columns for a distance of one sixth of the clear column height, but not less than 18 inches nor the maximum column dimension. Tie spacing for the remaining column height shall be not more than 16 bar diameters, 48 tie diameters or the least column dimension, but not more than 18 inches.

B. Lateral Tie Spacing, Seismic Zones 0, 1 and 2.

Lateral column ties around vertical bars for columns in seismic zones 0, 1 and 2 are illustrated in **Figure 4.47.**

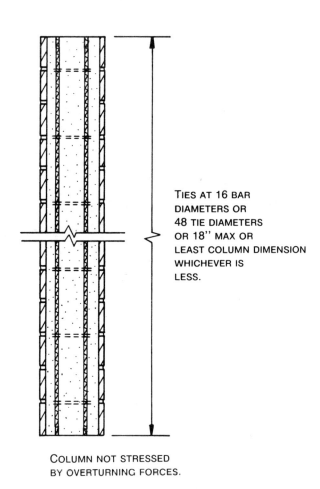

TIES AT 16 BAR DIAMETERS OR 48 TIE DIAMETERS OR 18" MAX OR LEAST COLUMN DIMENSION WHICHEVER IS LESS.

COLUMN NOT STRESSED BY OVERTURNING FORCES.

Figure 4.47 *Tie spacing in columns in seismic zones 0, 1, and 2.*

C. Laterial Tie Spacing, Seismic Zones 3 and 4.

The Uniform Building Code specifies additional requirements for seismic zones 3 and 4, as shown in **Figure 4.48**.

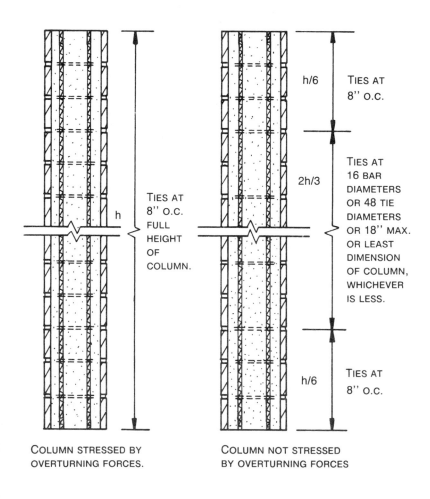

COLUMN STRESSED BY
OVERTURNING FORCES.

COLUMN NOT STRESSED
BY OVERTURNING FORCES

Figure 4.48 Tie spacing in columns in seismic zones 3 and 4.

D. Layout of Ties in concrete masonry columns
is shown in **Figure 4.49.**

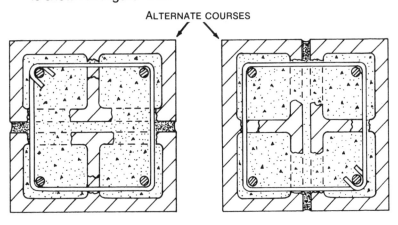

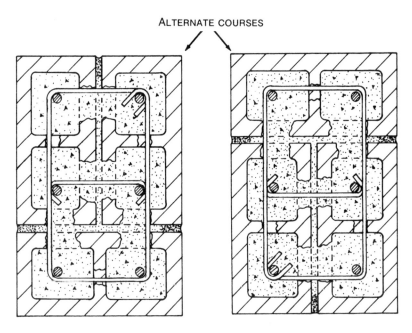

Figure 4.49 *Layout of concrete masonry units for column with tie details.*

ALTERNATE COURSES

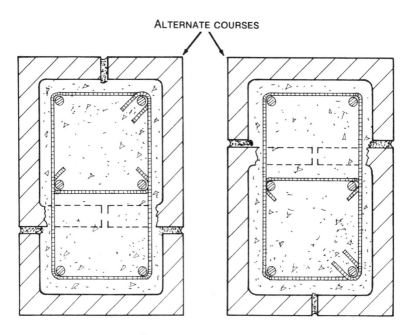

BUILT WITH PILASTER UNITS

ALTERNATE COURSES

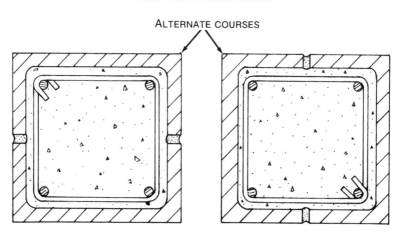

Figure 4.49 (Continued)

4.7.3 Projecting Wall Columns or Pilasters

Girders framing into a wall that are heavily loaded may require substantial base plates and columns to carry the load. Columns may be built projecting out from the wall to provide a convenient seat or surface to support the girders.

Projecting pilasters also serve to stiffen the wall and are supported at the top and bottom. The wall inbetween the pilasters can be designed to span horizontally. By this technique very high walls can be built using nominal masonry thicknesses. See **Figures** 4.50 and **4.51.**

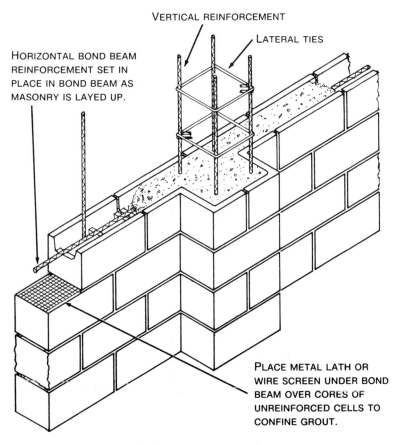

VERTICAL REINFORCEMENT

LATERAL TIES

HORIZONTAL BOND BEAM REINFORCEMENT SET IN PLACE IN BOND BEAM AS MASONRY IS LAYED UP.

PLACE METAL LATH OR WIRE SCREEN UNDER BOND BEAM OVER CORES OF UNREINFORCED CELLS TO CONFINE GROUT.

Figure 4.50 *Construction of reinforced concrete masonry pilaster with continuous bond beam.*

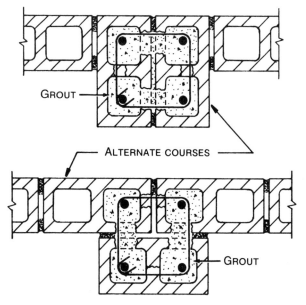

GROUT

ALTERNATE COURSES

GROUT

BUILT WITH TWO CORE STANDARD MASONRY UNITS.

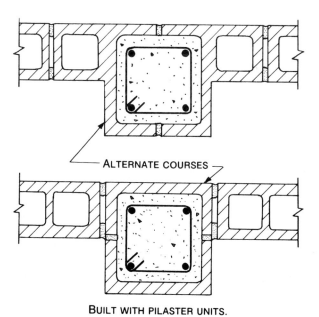

ALTERNATE COURSES

BUILT WITH PILASTER UNITS.

Figure 4.51 Projecting wall column masonry unit details.

4.7.4 Flush Wall Columns, Pilasters and Compression Steel at End of Walls

If engineering design permits, it is to the economic benefit of the owner and to the construction benefit of the contractor to build columns that are contained in the wall and are flush with the wall. The wall-contained columns permit faster construction, cause no projections from the wall, and do not require special units. The reinforcing steel must be tied in accordance with the code requirements (refer to **Figure 4.52**).

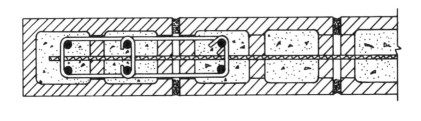

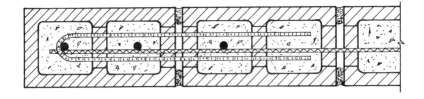

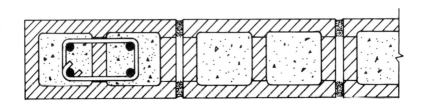

Figure 4.52 *Flush wall columns, ties around column bars and compression bars at end of wall.*

4.7.5   Ties on Compression Steel in Beams

See **Figure 4.53** for an illustration of ties for compression steel in beams.

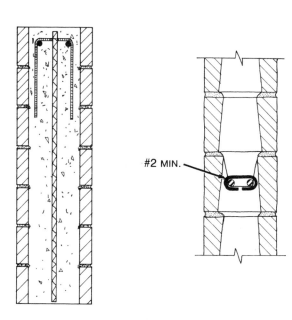

Figure 4.53 *Ties for compression steel in beams.*

4.7.6   Anchor Bolts

A. Anchor Bolt Clearance. It is necessary to have clearance around anchor bolts so that grout can fully surround the bolt. At least 1/4'' minimum clearance, 1/2'' preferred, shall be provided between the bolt and masonry unit.

> **U.B.C. Sec. 2407 (f)**
>
> (f) **Protection of Ties, Bolts and Joint Reinforcement.** A minimum of ⅝-inch mortar cover shall be provided between ties or joint reinforcement and any exposed face. The thickness of grout or mortar between masonry units and joint reinforcement or bolts shall be not less than ¼ inch, except that ¼ inch or smaller diameter reinforcement or bolts may be placed in bed joints which are at least twice the thickness of the reinforcement.

B. Anchor Bolt Ties. In order that lateral forces on anchor bolts be transferred to vertical steel, proper ties must be used around the anchor bolts and steel.

> **U.B.C. Sec. 2409 (b) 5. C.**
>
> C. **Anchor bolt ties.** Additional ties shall be provided around anchor bolts which are set in the top of the column. Such ties shall engage at least four bolts or, alternately, at least four vertical column bars or a combination of bolts and bars totaling four in number. Such ties shall be located within the top 5 inches of the column and shall provide a total of 0.4 square inch or more in cross-sectional area. The uppermost tie shall be within 2 inches of the top of the column. [See **Figure 4.54.**]

C. Anchor Bolts in Walls. Anchor bolts must be placed with adequate edge distance and spacing to ensure adequate performance. Refer to **Table 4-J** and **Figure 4.55.**

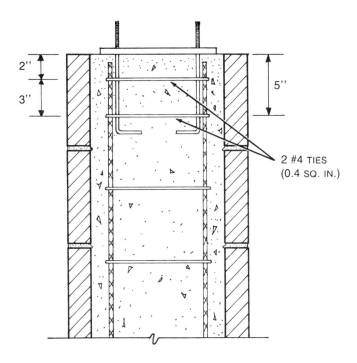

Figure 4.54 *Ties of anchor bars on top of columns.*

Table 4-J	Minimum Anchor Bolt Spacing and Edge Distance	
Bolt Dia. inches	Minimum	
	Spacing inches	Edge Distance inches
1/4	3.0	1.50
3/8	4.5	2.25
1/2	6.0	3.00
5/8	7.5	3.75
3/4	9.0	4.50
7/8	10.5	5.25
1	12.0	6.00
1-1/8	13.5	6.75

147

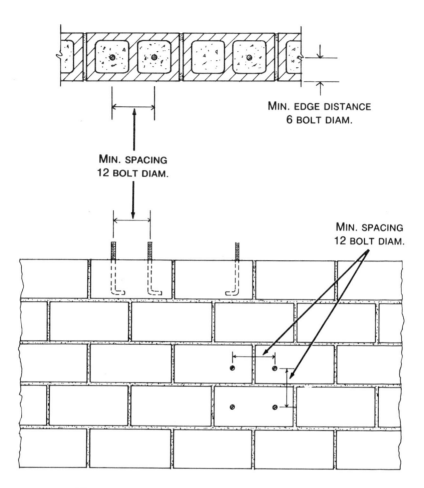

Figure 4.55 *Anchor bolt spacing and edge distance.*

D. Embedment of Anchor Bolts.

U.B.C. Sec. 2404 (f) 2.

2. **Construction requirements.** Reinforcement shall be placed prior to grouting. Bolts shall be accurately set with templates or by approved equivalent means and held in place to prevent movement.

U.B.C. Sec. 2406 (h) 1.

(h) **Shear and Tension on Embedded Anchor Bolts.**
1. General. A. Allowable loads and placement requirements for plate anchor bolts, headed anchor bolts and bent bar anchor bolts shall be determined in accordance with this section. The bent bar anchor bolt shall have a hook with a 90-degree bend with an inside diameter of three bolt diameters, plus an extension of 1½ bolt diameters at the free end.

B. The effective embedment length lb for plate or headed anchor bolts shall be the length of embedment measured perpendicular from the surface of the masonry to the bearing surface of the plate or head of the anchorage, and lb for bent bar anchors shall be the length of embedment measured perpendicular from the surface of the masonry to the bearing surface of the bent end minus one anchor bolt diameter. All bolts shall be grouted in place with at least 1 inch of grout between the bolt and the masonry.

U.B.C. Sec. 2406 (h) 5, 6.

5. **Minimum edge distance.** The minimum value of l_{be} measured from the edge of the masonry parallel with the anchor bolt to the surface of the anchor bolt shall be 1½ inches.

6. **Minimum embedment depth.** The minimum embedment depth shall be four bolt diameters.

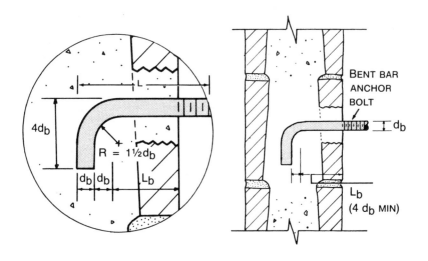

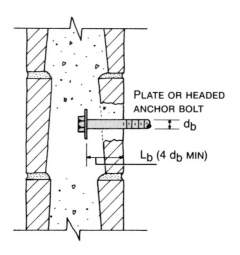

Figure 4.56 *Embedment of plate or headed anchor bolts and bent bar anchor bolts.*

The Golden Nugget Towers *in Las Vegas, 22 stories of load-bearing concrete masonry. (Architect, Joel D. Bergman, AIA; Project Architect, DeRuyter O. Butler. Photo by Allen Photographic Services, Inc.)*

151

U.B.C. TABLE NO. 24-D-1—ALLOWABLE TENSION, B_t, ON BENT BAR ANCHOR BOLTS FOR CLAY AND CONCRETE MASONRY, pounds[1][2][3]

f'_m (psi)	EMBEDMENT LENGTH, l_b, or EDGE DISTANCE, l_{be}, inches						
	2	3	4	5	6	8	10
1500	240	550	970	1520	2190	3890	6080
1800	270	600	1070	1670	2400	4260	6660
2000	280	630	1120	1760	2520	4500	7020
2500	310	710	1260	1960	2830	5030	7850
3000	340	770	1380	2150	3100	5510	8600
4000	400	890	1590	2480	3580	6360	9930
5000	440	1000	1780	2780	4000	7110	11,100
6000	480	1090	1950	3040	4380	7790	12,200

[1] The allowable tension values in Table No. 24-D-1 are based on compressive strength of masonry assemblages. Where yield strength of anchor bolt steel governs, the allowable tension in pounds is given in Table No. 24-D-2.

[2] Values are for bolts of at least A 307 quality. Bolts shall be those specified in Section 2406(h) 1 A.

[3] Values shown are for work with or without special inspection.

U.B.C. TABLE NO. 24-D-2—ALLOWABLE TENSION, B_t, ON BENT BAR ANCHOR BOLTS FOR CLAY AND CONCRETE MASONRY, pounds[1][2]

BENT BAR ANCHOR BOLT DIAMETER, inches						
3/8	1/2	5/8	3/4	7/8	1	1 1/8
790	1410	2210	3180	4330	5650	7160

[1] Values are for bolts of at least A 307 quality. Bolts shall be those specified in Section 2406(h) 1 A.

[2] Values shown are for work with or without special inspection.

U.B.C. TABLE NO. 24-E—ALLOWABLE SHEAR, B_v, ON BENT BAR ANCHOR BOLTS FOR CLAY AND CONCRETE MASONRY, pounds[1][2]

f'_m (psi)	BENT BAR ANCHOR BOLT DIAMETER, inches						
	3/8	1/2	5/8	3/4	7/8	1	1 1/8
1500	480	850	1330	1780	1920	2050	2170
1800	480	850	1330	1860	2010	2150	2280
2000	480	850	1330	1900	2060	2200	2340
2500	480	850	1330	1900	2180	2330	2470
3000	480	850	1330	1900	2280	2440	2590
4000	480	850	1330	1900	2450	2620	2780
5000	480	850	1330	1900	2590	2770	2940
6000	480	850	1330	1900	2600	2900	3080

[1] Values are for bolts of at least A 307 quality. Bolts shall be those specified in Section 2406(h) 1 A.
[2] Values shown are for work with or without special inspection.

4.8 SPECIAL PROVISIONS FOR SEISMIC DESIGN AND CONSTRUCTION

4.8.1 General

Earthquakes are a recognized threat to life, safety and buildings. To help prevent loss of life and reduce damage to structures, special detailing requirements are imposed by the Uniform Building Code. These requirements are based on the seismic zone in which the building is located.

U.B.C. Sec. 2407 (h) 1.

(h) **Special Provisions in Areas of Seismic Risk.** 1. **General.** Masonry structures constructed in the seismic zones shown in Figure No. 2 of Chapter 23 shall be designed in accordance with the design requirements of this chapter and the special provisions for each seismic zone given in this section.

U.B.C. Sec. 2312 (h) 2. H.

H. Anchorage of concrete or masonry walls. Concrete or masonry walls shall be anchored to all floors and roofs which provide lateral support for the wall. The anchorage shall provide a positive direct connection between the wall and floor or roof construction capable of resisting the horizontal forces specified in Section 2312 (g) or Section 2310. Requirements for developing anchorage forces in diaphragms are given in Section 2312 (h) 2 I below. Diaphragm deformation shall be considered in the design of the supported walls.

U.B.C. Sec. 2312 (h) 2. H.I. (iv)

(iv) Where wood diaphragms are used to laterally support concrete or masonry walls, the anchorage shall conform to Section 2312 (h) 2 H above. In Seismic Zones Nos. 3 and 4 anchorage shall not be accomplished by use of toe nails or nails subject to withdrawal, nor shall wood ledgers or framing be used in cross-grain bending or cross-grain tension, and the continuous ties required by paragraph (iii) above shall be in addition to the diaphragm sheathing.

4.8.2 Seismic Zones Nos. 0 and 1

U.B.C. Sec. 2407 (h) 2.

2. Special provisions for Seismic Zones Nos. 0 and 1. There are no special design and construction provisions in this section for structures built in Seismic Zones Nos. 0 and 1.

Horizontal reinforcement should be placed in the top of footings, on all sides of wall openings, at roof and floor levels and at the top of parapet walls. Horizontal reinforcement in a wall is generally

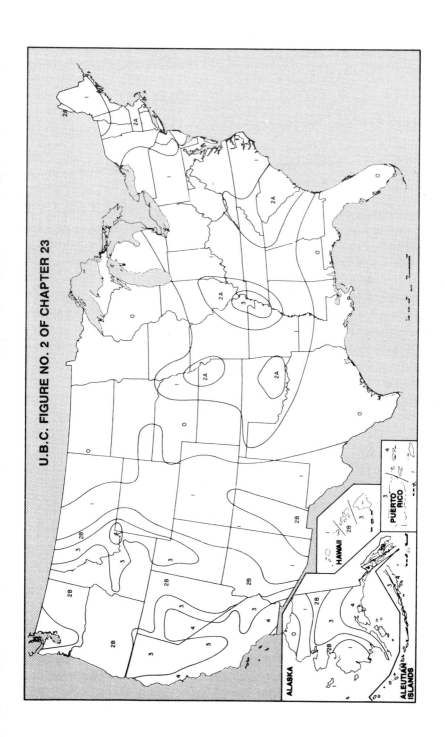

U.B.C. FIGURE NO. 2 OF CHAPTER 23

155

for temperature and shrinkage requirements, and the exact position-ing is not critical. It can be placed into the fresh grout or laid on top of the cross webs in concrete masonry. The horizontal steel should be placed with enough clearance between bars to allow the grout to flow through and completely surround the bar. This will help in-sure the development of good bond.

4.8.3 Seismic Zone No. 2

U.B.C. Sec. 2407 (h) 3.

3. **Special provisions for Seismic Zone No. 2.** Masonry structures in Seismic Zone No. 2 shall comply with the following special provisions.

A. **Materials.** The following materials shall not be used as part of the structural frame: Type O mortar, masonry cement, plastic cement, nonload-bearing masonry units and glass block.

U.B.C. Sec. 2407 (h) 3. B.

B. **Wall reinforcement.** Vertical reinforcement of at least 0.20 square inch in cross-sectional area shall be provided continuously from support to support at each corner, at each side of each opening, at the ends of walls and at a maximum spacing of 4 feet apart, horizontally through-out the wall.

Horizontal reinforcement not less than 0.2 square inch in cross-sectional area shall be provided: (1) at the bottom and top of wall openings and shall extend not less than 24 inches nor less than 40 bar diameters past the opening, (2) continuously at structurally connected roof and floor levels and at the top of walls, (3) at the bottom of the wall or in the top of the foundations when dowelled to the wall, (4) at maximum spacing of 10 feet unless uniformly

Reinforcement at the top and bottom of openings when continuous in the wall may be used in determining the maximum spacing specified in Item No. (1) above.

C. Stack bond. Where stack bond is used, the minimum horizontal reinforcement ratio shall be .007*bt*. This ratio shall be satisfied by uniformly distributed joint reinforcement or by horizontal reinforcement spaced not over 4 feet and fully embedded in grout or mortar.

Figure 4.57 depicts minimum reinforcing bar locations in Seismic Zone 2.

4.8.4 Seismic Zones Nos. 3 and 4

U.B.C. Sec. 2407 (h) 4.

4. Special provisions for Seismic Zones Nos. 3 and 4. All masonry structures built in Seismic Zones Nos. 3 and 4 shall be designed and constructed in accordance with requirements for Seismic Zone No. 2 and with the following additional requirements and limitations.

> **EXCEPTION:** One- and two-story structures of Group R, Division 3 and Group M occupancies in Seismic Zone No. 3 with h'/t not greater than 27 and using running bond construction may be constructed in accordance with the requirements of Seismic Zone No. 2.

The f'_m from Table No. 24-C shall be limited to a maximum of 1500 psi for concrete masonry and 2600 psi for clay masonry unless f'_m is verified by prism tests as required in Section 2405 (c) 1.

Reinforced hollow-unit stacked bond construction which is part of the seismic resisting system shall use open-end units so that all head joints are made solid, shall use bond-beam units to facilitate the flow of grout and shall be grouted solid.

A. Materials. The following materials shall not be used as part of the structural frame: Type N mortar.

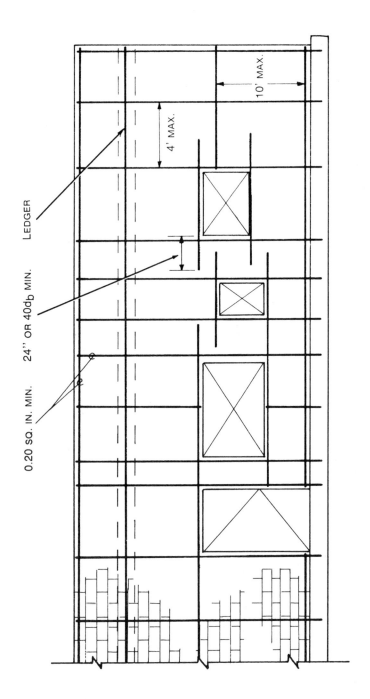

Figure 4.57 *Minimum reinforcing for seismic zone No. 2.*

LEDGER

24" OR 40d$_b$ MIN.

0.20 SQ. IN. MIN.

4' MAX.

10' MAX.

158

B. **Wall reinforcement.** All walls shall be reinforced with both vertical and horizontal reinforcement. The sum of the areas of horizontal and vertical reinforcement shall be at least 0.002 times the gross cross-sectional area of the wall, and the minimum area of reinforcement in either direction shall not be less than 0.0007 times the gross cross-sectional area of the wall. The spacing of reinforcement shall not exceed 4 feet. The diameter of reinforcement shall not be less than ⅜ inch except that joint reinforcement may be considered as part or all of the requirement for minimum reinforcement. Reinforcement shall be continuous around wall corners and through intersections. Only horizontal reinforcement which is continuous in the wall or element shall be considered in computing the minimum area of reinforcement. Reinforcement with splices conforming to Section 2409 (e) 6 shall be considered as continuous reinforcement.

C. **Column reinforcement.** The spacing of column ties shall not be more than: 8 inches the full height for columns stressed by tensile or compressive axial overturning forces due to the seismic loads of Section 2312; 8 inches for the tops and bottoms of all other columns for a distance of one sixth of the clear column height, but not less than 18 inches nor the maximum column dimension. Tie spacing for the remaining column height shall be not more than 16 bar diameters, 48 tie diameters or the least column dimension, but not more than 18 inches. [See Figure 4.44 for details.]

U.B.C. Sec. 2407 (h) 4.

D. **Stack bond.** Where stack bond is used, the minimum horizontal reinforcement ratio shall be .0015 bt. If open-end units are used and grouted solid, then the minimum horizontal reinforcement ratio shall be .0007 bt.

E. **Minimum dimension.** (i) **Bearing walls.** The nominal thickness of reinforced masonry bearing walls shall not be less than 6 inches except that nominal 4-inch-thick load-bearing reinforced hollow clay unit masonry walls may be used, provided net area unit strength exceeds 8000 psi, units are laid in running bond, bar sizes do not exceed ½ inch with no more than two bars or one splice in a cell, and joints are flush cut, concave or a protruding V section. Minimum bar coverage where exposed to weather shall be 1½ inches.

(ii) **Columns.** The least nominal dimension of a reinforced masonry column shall be 12 inches except that if the allowable stresses are reduced to one half the values given in Section 2406, the minimum nominal dimension shall be 8 inches.

Seismic Zones 3 and 4 also require that reinforcing be continuous around corners. The sum of the areas of horizontal and vertical reinforcements shall be at least 0.002 times the gross cross-sectional area of the wall, and the minimum area of reinforcement in either direction shall not be less than 0.0007 times the gross cross-sectional area of the wall.

Figure 4.58 depicts minimum reinforcing bar locations in Seismic Zones 3 and 4.

4.8.5 Shear Walls

U.B.C. Sec. 2407 (h) 4. F. (ii)

(ii) **Reinforcement.** The portion of the reinforcement required to resist shear shall be uniformly distributed and shall be joint reinforcing, deformed bars, or a combination thereof. The maximum spacing of reinforcement in each direction shall be not less than the smaller of one half the length or height of the element nor more than 48 inches. [See **Figure 4.38.**]

Joint reinforcement used in exterior walls and considered in the determination of the shear strength of the member shall be hot-dipped galvanized in accordance with U.B.C. Standard No. 24-15.

Reinforcement required to resist in-plane shear shall be terminated with a standard hook or with an extension of

In bearing walls of every type of reinforced masonry there shall be not less than a #4 bar or two #3 bars on all sides of, and adjacent to, every opening which exceeds 24 inches in either direction, and such bars shall extend not less than 40 bar diameters, but in no case less than 24 inches, beyond the corners of the opening. These bars are required in addition to the minimum reinforcement.

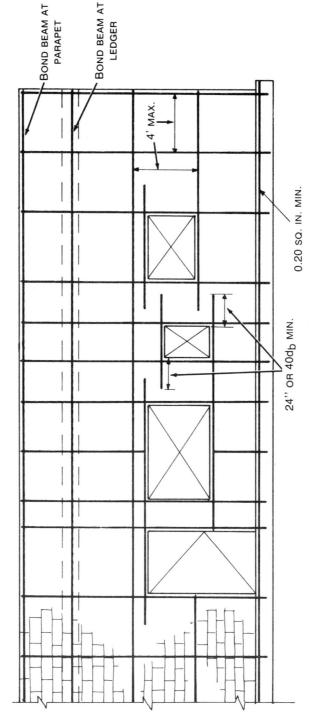

Figure 4.58 Minimum reinforcing bar locations in seismic zones 3 and 4.

161

proper embedment length beyond the reinforcing at the end of the wall section. The hook or extension may be turned up, down or horizontally. Provisions shall be made not to obstruct grout placement. Wall reinforcement terminating in columns or beams shall be fully anchored into these elements.

U.B.C. Sec. 2407 (h) 4. G.

G. **Hooks.** The term "hook" or "standard hook" as used herein for tie anchorage in Seismic Zones No. 3 and No. 4 shall mean a minimum turn of 135 degrees plus an extension of at least six bar diameters, but not less than 4 inches at the free end of the bar. [See Figure 4.41.]

> **EXCEPTION:** Where the ties are placed in the horizontal bed joints, the hook shall consist of a 90-degree bend having a radius of not less than four bar diameters plus an extension of 32 bar diameters. [See Figure 4.42.]

H. **Mortar joints between masonry and concrete.** Concrete abutting structural masonry such as at starter courses or at wall intersections not designed as true separation joints shall be roughened to a full amplitude of 1/16 inch and shall be bonded to the masonry per the requirements of this chapter as if it were masonry. Unless keys or proper reinforcement are provided, vertical joints as per Section 2407 (b) 2 shall be considered to be stack bond and the reinforcement as required for stack bond shall extend through the joint and be anchored into the concrete.

4.9 GROUTING OF CONCRETE MASONRY WALLS

4.9.1 General

Section 2404 (f) of the Uniform Building Code states the allowable grouting requirements as outlined by Table No. 24-G.

U.B.C. Sec. 2404 (f) 1.

(f) **Grouted Masonry. 1. General conditions.** Grouted masonry shall be constructed in such a manner that all elements of the masonry act together as a structural element.

4.9.2 Mortar Protrusions

U.B.C. Sec. 2404 (f). 1.

Prior to grouting, the grout space shall be clean so that all spaces to be filled with grout do not contain mortar projections greater than ½ inch, mortar droppings or other foreign material. [See **Figure 4.59**.]

Mortar projections should not obstruct the placement and consolidation of grout. Reasonable care shall be taken either to prevent

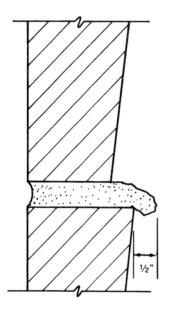

½"

Figure 4.59
*Mortar projection
into grout space.*

mortar projections as the masonry units are being placed or the projections shall be removed while the mortar is plastic or broken off when hard and removed from the cleanout openings.

4.9.3 Grout Slump

Grout shall be plastic with a slump fluidity of 8'' to 10'' and be cohesive to avoid segregation of pea gravel. See **Figure 4.60.**

U.B.C. Sec. 2404 (f) 1.

Grout shall be placed so that all spaces designated to be grouted shall be filled with grout and the grout shall be confined to those specific spaces.

Grout materials and water content shall be controlled to provide adequate fluidity for placement, without segregation of the constituents and shall be mixed thoroughly.

4.9.4 Grouting Limitations

U.B.C. Sec. 2404 (f) 1.

The grouting of any section of wall shall be completed in one day with no interruptions greater than one hour.

Between grout pours, a horizontal construction joint shall be formed by stopping all wythes at the same elevation and with the grout stopping a minimum of 1½ inches below a mortar joint, except at top of wall. Where bond beams occur, stop grout pour a minimum of ½ inch below the top of the masonry.

Size and height limitations of the grout space or cell shall not be less than shown in Table No. 24–G. Higher grout pours or smaller cavity widths or cell size than shown in Table No. 24–G may be used when approved, if it is demonstrated that grout spaces are properly filled.

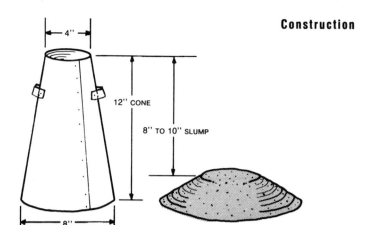

Figure 4.60 *Slump of grout 8" to 10".*

U.B.C. TABLE NO. 24-G—GROUTING LIMITATIONS

GROUT TYPE	GROUT POUR[2] MAXIMUM HEIGHT (Feet)	LEAST CLEAR DIMENSIONS[1]		CLEANOUTS REQUIRED[4]
		Width of Grout Space[5] (In.)[3]	Cell Dimensions[6] (In. x In.)	
Fine	1	¾	1½ × 2	No
Fine	5	1½	1½ × 2	No
Fine	8	1½	1½ × 3	Yes
Fine	12	1½	1¾ × 3	Yes
Fine	24	2	3 × 3	Yes
Coarse	1	1½	1½ × 3	No
Coarse	5	2	2½ × 3	No
Coarse	8	2	3 × 3	Yes
Coarse	12	2½	3 × 3	Yes
Coarse	24	3	3 × 4	Yes

[1] The clear dimension is the cell or grout space width less mortar projections.
[2] For grout pours over 5 feet high, see Section 2404(f) 1.
[3] Grout space width shall be increased by the horizontal projection of the diameters of the horizontal bars within the cross section of the grout space.
[4] Cleanouts may be omitted if approved provisions are made to keep the grout space clean prior to grouting.
[5] For grout spaces in grouted multiwythe masonry.
[6] For grout cells in grouted hollow unit masonry.

Section 2.11.2 of this handbook explains the types of grout, fine and coarse.

4.9.5 Low Lift Grouting

The grouting method commonly known as low lift grouting is done in grout pours of five feet or less in height.

The wall is constructed in five foot increments. Prior to grouting, horizontal and vertical reinforcement bolts and other embedded items are positioned. It is important that sufficient time is allowed for the mortar joints to set and be able to withstand the grout pressure.

Grout is poured into all reinforced cells, and other designated cells if required, to a height approximately 1-1/2'' above or below the last mortared bed joint. The last lift is poured to the top of the wall.

Vertical cells to be filled must align vertically to maintain a continuous unobstructed cell area not less than 1-1/2 inches x 2 inches. Horizontal beams to be grouted should be isolated horizontally with metal lath or special concrete block units to prevent the grout from flowing into cells that should be void. Paper should not be used for this purpose, however, 30-pound felt may be used if it does not extend into the mortar joint.

The principal advantage of the low-lift grouting method is that cleanouts or inspection openings are not required. The inspector can visually check the cells for proper alignment, check that the bottom of the cells are clean and free of excessive mortar protrusions, and verify the reinforcing steel location, all before grouting the wall.

Between grout pours, a horizontal construction joint shall be formed by stopping all wythes at the same elevation and with the grout stopping a minimum of 1-1/2'' below a mortar joint, except at the top of a wall. Where bond beams occur, stop the grout pour a minimum of 1/2'' below the top of the masonry. Refer to **Figure 4.61**.

When cleanouts are not used (low lift grouting), the height of any grout pour shall not exceed five feet. If the grout pours are only 12 inches or less in height, the grout may be consolidated by puddling.

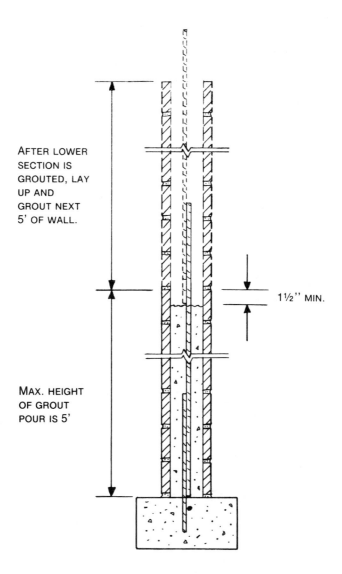

AFTER LOWER
SECTION IS
GROUTED, LAY
UP AND
GROUT NEXT
5' OF WALL.

1½'' MIN.

MAX. HEIGHT
OF GROUT
POUR IS 5'

DELAY APPROXIMATLY 3 TO 5 MINUTES ALLOWING THE WATER
TO BE ABSORBED BY THE MASONRY UNITS, THEN CONSOLIDATE
THE GROUT BY MECHANICALLY VIBRATING.

Figure 4.61 Grouting without cleanouts, commonly called
low-lift grouting.

For grout pours greater than 12", the grout must be consolidated by mechanical vibration and reconsolidated by mechanical vibration prior to the grout losing its plasticity.

U.B.C. Sec. 2404 (f) 1.

The grouting of any section of wall shall be completed in one day with no interruptions greater than one hour.

4.9.6 Cleanouts

A cleanout is an opening or hole of sufficient size through the face of the block used to successfully clean out all mortar droppings and other debris from the bottom of the cell that is to be grouted. Cleanouts are illustrated in **Figure 4.62.**

U.B.C. Sec. 2404 (f) 1.

When required by Table No. 24–G, cleanouts shall be provided in the bottom course at every vertical bar but shall not be spaced more than 32 inches on center for solidly grouted masonry. When cleanouts are required, they shall be sealed after inspection and before grouting.

4.9.7 High-Lift Grouting

A. Uniform Building Code Requirements. When grout pours exceed five feet, cleanouts must be provided and the method commonly known as high-lift grouting must be used.

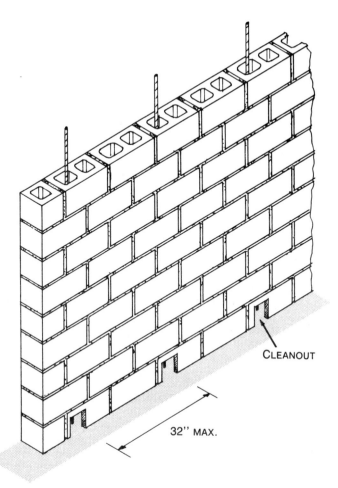

Figure 4.62 *Cleanouts in wall for grout pours higher than five feet.*

In high lift grouting, illustrated in **Figure 4.63**, the walls are built to their full height before grouting, up to a maximum of 24 feet. Cleanout holes are required at the bottom of all vertical cells containing reinforcing, but not more than 32 inches apart for solid grouted masonry. It is recommended that for walls that are solid grouted the first course be an inverted bond beam unit to allow for cleaning mortar droppings or debris from the foundation and between cleanouts, which may be as much as 32" o.c. This also improves the flow of grout at the foundation.

> **U.B.C. Sec. 2404 (f) 1.**
>
> Units may be laid to the full height of the grout pour and grout shall be placed in a continuous pour in grout lifts not exceeding 6 feet.

Mortar projections exceeding the vertical height of the mortar joint, or 1/2'', and all mortar droppings shall be cleaned out of the grout cells and off the reinforcing steel prior to grouting. ''Clean'' does not mean ''surgically clean,'' but merely no loose deleterious material in the areas to be grouted.

There are various methods of cleaning mortar droppings and overhangs which include using compressed air, a rod, a stick, or a high-pressure jet stream of water to dislodge the material.

Grout should not be poured until the mortar has set a sufficient time to adequately withstand the pressure of the grout.

All reinforcing steel, bolts, other embedded items and cleanout closures shall be properly secured in place before grouting and should be inspected during grouting.

Unless otherwise indicated, the grout should have an 8'' to 10'' slump.

Grout is placed in lifts not exceeding six feet and consolidated at the time of placing by a mechanical vibrator. After each lift is placed, wait for absorption of water into the block, approximately 5 to 10 minutes, and then reconsolidate the grout before it loses its plasticity. Place the next lift immediately, or as soon as reasonable. The full height of any section of wall should be completed in one day, with no interruption between lifts greater than one hour. Reconsolidate the last lift at the top of the wall and fill the grout space to the top.

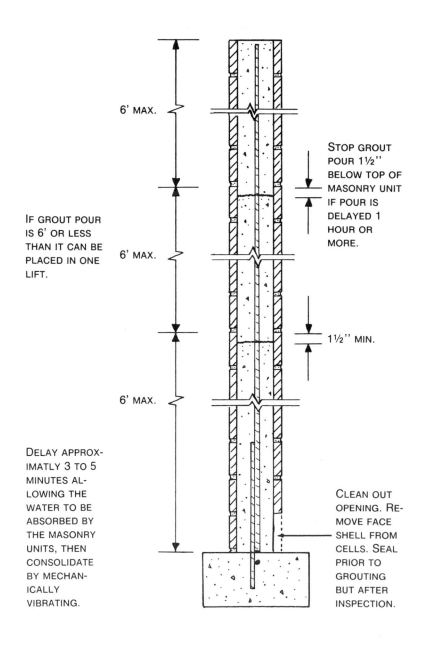

6' MAX.

IF GROUT POUR
IS 6' OR LESS
THAN IT CAN BE
PLACED IN ONE
LIFT.

6' MAX.

STOP GROUT
POUR 1½"
BELOW TOP OF
MASONRY UNIT
IF POUR IS
DELAYED 1
HOUR OR
MORE.

1½" MIN.

6' MAX.

DELAY APPROX-
IMATLY 3 TO 5
MINUTES AL-
LOWING THE
WATER TO BE
ABSORBED BY
THE MASONRY
UNITS, THEN
CONSOLIDATE
BY MECHAN-
ICALLY
VIBRATING.

CLEAN OUT
OPENING. RE-
MOVE FACE
SHELL FROM
CELLS. SEAL
PRIOR TO
GROUTING
BUT AFTER
INSPECTION.

Figure 4.63 High-lift method of grouting block.

B. Office of the State Architect—IR 24–4 Requirements
Rev. 1/86 Ref. Sec. 2-2415, Title 24

FILLED CELL CONCRETE MASONRY
HIGH LIFT GROUTING METHOD

(a) **General.** Title 24, CAC requires that where high lifting is used the method is to be approved by the Office of the State Architect. The procedure described herein is an acceptable method.

Except as noted otherwise herein, the requirements of Chapter 2-24, Title 24 govern.

(b) **Description.** The High Lift Grouting Method as developed for use in reinforced concrete block masonry is intended for use on wall construction where openings, block pattern arrangements, special reinforcing steel, or embedded structural steel details do not prevent the free flow of grout or inhibit the use of mechanical vibration to properly consolidate the grout fill in all cells or horizontal grout spaces. Horizontal reinforcing should be positioned in a single vertical plane at each curtain of steel to allow maximum accessibility to the cell spaces.

The procedure requires that all masonry units, reinforcing steel and embedded items will be in place before grouting of the wall voids commences. The work should be so arranged that once grouting of a section of wall is started the grouting is to proceed in lifts without stopping except as noted below until the full height of the prepared section is poured. The waiting period between lifts is to be limited to the time required to obtain an initial consolidation of grout due to settlement shrinkage and absorption of excess water by the masonry units. This also allows for a reduction in hydrostatic pressure of the grout on the masonry unit and reduces the possibility of "blowouts".

The grout shall be a high slump workable mix preferably placed by pumping to permit continuous pouring and is to be worked into all voids. Use mechanical vibrators for consolidation. Where job conditions preclude such use other methods may be employed if approved by the Office of the State Architect. Because of the high water-cement ratio used in this type of grout it is essential that the grout be reconsolidated after it has taken on a plastic consistency but prior to taking an initial set. The reconsolidation is intended to overcome settlement shrinkage separations from the reinforcing steel and to promote bond to the masonry unit walls.

For the purose of this IR a "pour" is considered as the entire height of grout fill placed in one day and is composed of a number of successively placed grout lifts. A "lift" is the layer of grout placed in a single continuous operation.

The maximum height of pour is limited by the practical considerations of segregation of grout due to the height of free fall, effect of dry grout deposits left on block projections and reinforcing steel and the ability to effectively reconsolidate the grout. Unless specifically approved otherwise the maximum height of pour will be twelve feet for eight-inch walls and sixteen feet for twelve inch walls. For height of lift see Section (e) (9).

(c) **Quality of Materials.** All materials are to conform to Section 2403, UBC and Section 2-2403, Title 24 with the following additional requirements:

(1) Pea Gravel. Pea gravel for grout is to conform to UBC Standard 24–23, and Table 24–23–A, Coarse Aggregate, except when other gradings are specifically approved by the architect or structural engineer, and the Office of the State Architect.

(2) Coarse Aggregate. Coarse aggregate is to conform to Section 2-2603, Title 24.

(3) Admixture. Use an approved grout admixture of a type that reduces early water loss to the masonry units and produces an expansive action in the plastic grout sufficient to offset initial shrinkage and promote bonding of the grout to all interior surfaces of the masonry units. Obtain approval for use of the admixture from the architect or structural engineer and the Office of the State Architect. (See IR 24-2 for types of admixtures.)

(d) **Mortar and Grout.**

(1) Mortar. Mortar is to comply with the requirements of Section 2-2403(r), Title 24 and with the following additional requirements:

(A) Place approximately half the required water and sand into the mixer while running.

(B) Add cement and the remainder of the sand and water into the mixer in that order and mix for a period of at least two minutes.

(C) Add lime and continue mixing as long as needed to secure a uniform mass.

(D) The total mixing time may not be less than 10 minutes.

(2) Grout. The grout mix is to comply with the requirements of Section 2-2403(s), Title 24.

Sufficient water may be added to make a workable mix that will flow into all joints of the masonry without separation or segregation. When grout is to be placed in masonry units with typical rates of absorption the slump of the grout should be approximately 9 to 10 inches depending on temperature and humidity conditions.

Where the least lateral dimension of cells to be filled exceeds five inches a coarser aggregate may be used in the grout fill if the mix is designed in accordance with Section 2-2604, Title 24. The maximum size of aggregate is not to exceed one inch. The water per sack of cement may be greater than is shown in Table 2-26-A, Title 24 to allow for absorption by the masonry units and with sufficient workability to meet the requirements given in the paragraph above.

Grout mixes are to contain an approved admixture conforming to the requirements of paragraph (c) (3) above and IR 24-2. Use such admixture in accordance with manufacturer's instructions.

(3) Mixing of Grout. The mixing of grout is to conform to the requirements for mixing of concrete, Section 2-2605(b) of Title 24.

Whenever possible mix and deliver grout in accordance with the requirements for transit-mixed concrete.

Time the addition of the admixture in strict accordance with the manufacturer's instructions. The procedure used for adding it to the grout mix is to provide for good dispersion.

(4) Certification. The quality and quantities of materials used in transit-mixed grout are to be continuously checked by a qualified person at the location where the materials are measured.

If specified by the architect or structural engineer and approved by the Office of the State Architect, certification concerning quantity of materials may be accepted from a licensed weighmaster in lieu of continuous plant inspection if the following procedures are used to check the quality of the materials to be used in the grout.

> (A) Test samples of the aggregate to be used in the grout are to be taken and tested by the testing laboratory in accordance with U.B.C. Standard 24-23.
>
> (B) The transit-mixed grout supplier uses a mix design for the proportions of cement, sand, and pea gravel or coarser aggregate prepared or approved by the project architect or structural engineer.
>
> (C) On the first half-day, transit-mixed grout is supplied to the job, and at such other times as may be required by the architect and/or structural engineer, and the quantity and quality of materials used in the transit-mixed grout are continuously checked by an approved inspector at the batch plant location. In addition to the quality of the aggregates, the inspector is to verify the quality of the cement.
>
> (D) The licensed weighmaster will certify to each load on a load ticket transmitted to the owners' inspector and furnish an affidavit at the completion of the project, all in accordance with OSA IR 26-7.

(5) Tests. Testing of mortar and grout is to conform to the requirements of Section 2-2404(d) of Title 24.

(e) **Construction.** The construction of high lift concrete block masonry work is to conform to the requirements of Chapter 2-24 of Title 24, with the following additional requirements:

(1) Foundations. The contact surface of all foundations and floors that are to receive masonry work are to be thoroughly cleaned and roughened in accordance with Section 2-2606(d), Title 24 before start of laying. Protect the roughened surface during construction to assure a good bond between the grout fill and the concrete surface.

(2) Cleanouts. Provide cleanout openings through block faces for all cells at the bottom of each pour. The openings are to be made prior to the start of laying and be of sufficient size and location to allow thorough flushing away of all mortar, droppings and debris.

After the laying of masonry units is completed, the cells cleaned, the reinforcing positioned and inspection completed, close the cleanouts by inserting face shells of masonry units or covering the openings with forms. Face shell plugs are to have a two-day minimum curing time and be adequately braced during grouting to resist the pressure of the fluid grout.

(3) Reinforcement. Place all reinforcing steel accurately in strict accordance with the approved plans and specifications. Both horizontal and vertical reinforcing are to be held in position by wire ties or spacing devices near ends and at intervals not exceeding 192 diameters of the reinforcement. Place the horizontal reinforcing as the work progresses. The vertical reinforcing may be dropped into position after the completion of the laying if adequate positioning devices are provided to hold the reinforcement in proper location.

(4) Masonry Units. Use of open end concrete masonry units is preferred wherever possible and is required for stacked bond. Bond beam units are to be used wherever possible to facilitate the horizontal flow of grout and are required at all horizontal bars to provide a minimum vertical opening at all cross webs three inches high by three inches wide.

The concrete masonry units need not be wetted before laying, except in dry areas, where the contact surfaces of the units should be moistened immediately before laying to prevent excessive drying of mortar.

(5) Laying. Fill all head and bed joints solidly with mortar for a distance in from the face of the unit not less than the thickness of the face shell. Care is to be taken in placing the mortar to keep a minimum of droppings from falling into the block cells. Arrange open end concrete masonry units used in stacked bond so the closed ends are not abutting.

(6) Wall Ties and Bracing. When stacked bond is used or when adequate cross webs between face shells are not provided, ties of heavy gauge wire embedded in the horizontal mortar joints should be provided across continuous vertical joints or between face shells to prevent "blow-outs" due to the hydrostatic pressure of the fluid grout. External ties or braces may also be used for this purpose.

During construction, brace the ungrouted walls adequately to resist wind and other forces.

(7) Mortar Droppings and Overhangs. Thoroughly remove all mortar droppings and overhangs from the foundation or bearing surface, cell walls and reinforcing. Acceptable methods for this are by hosing with a jet stream at least twice a day (at midday and quitting time) or by providing a two- or three-inch blanket of dry sand over the exposed surface of the foundation, dislodging any hardened mortar from the cell walls and reinforcing with a pole or rod and removing the mortar debris with the sand cover prior to clean up and grouting.

(8) Construction Joints. In the High Lift Grouting Method, intermediate horizontal construction joints are not permitted. Plan the work for one continuous pour of grout to the top of the wall in four foot layers or lifts in the same working day. Should a blow-out, a breakdown in equipment, or any other emergency occur, cease the grouting operation. An alternate procedure may be used with the approval of the architect or structural engineer and the Office of the State Architect.

The section of wall to be grouted in any one pour should be limited to a length in which successive lifts can be placed within one hour of the preceding lifts. Vertical control barriers shall be placed between pour sections in locations approved by the architect or structural engineer and the Office of the State Architect.

(9) Grouting. To prevent "blow-outs," pour no grout until the mortar has set and cured. However, grout the walls as soon as possible after mortar has cured to reduce shrinkage cracking of the vertical joints. All cleanout closures, reinforcing, bolts and embedded connection items are to be in position before grouting is started.

Handle grout from the mixer to the point of deposit in the grout space as rapidly as practical by pumping and placing methods which will prevent segregation of the mix and cause a minimum of grout splatter on reinforcing and masonry unit surfaces not being immediately encased in the grout lift. Depending upon weather conditions and absorption rates of the masonry units, the lift heights and waiting periods may be varied. Under normal weather conditions, with typical masonry units, the individual lifts of grout are limited to four feet in height with a waiting period between lifts of thirty to sixty minutes.

Place the first lift of grout to a uniform height within the pour section and mechanically vibrate thoroughly to fill all voids. The grouting team should be organized to enable the vibration to follow closely behind and at the same pace as the pouring operation.

After a waiting period sufficient to permit the grout to become plastic but before it has taken any set, the succeeding lift should be poured and alternate cells vibrated twelve inches to eighteen inches into the preceding lift. Do this in such a manner as to reconsolidate the preceding lift and close any plastic shrinkage cracks or separations from the cell walls.

If, because of unavoidable job conditions, the placing of the succeeding lift is going to be delayed beyond the period of workability of the preceding lift, reconsolidate each lift by re-working with the mechanical vibrator as soon as the grout has taken its settlement shrinkage.

Repeat the waiting, pouring and reconsolidation steps until the top of the pour is reached. Reconsolidate the top lift also after the required waiting period to fill any space left by settlement shrinkage.

(10) Cleaning Wall. Immediately after the wall has been fully grouted, hose off with water under pressure through a jet nozzle, all the scum and stains which have percolated through the blocks and joints.

(11) Curing. Attention should be given to proper curing of the mortar joints as well as the grout concrete pour. The concrete block work and top of grout pour should be kept damp to prevent too rapid drying during hot or drying weather, and drying winds.

(f) **Inspection and Core Tests.**

(1) Inspection. All masonry work is required to be continuously inspected during laying and grouting by an inspector specially approved for that purpose by the Office of the State Architect. The inspector makes test samples and performs such tests as are required by paragraph (d) (5) above.

The special masonry inspector is to check the materials, details of construction, and construction procedure. He will furnish a verified report that, of his own personal knowledge, the work covered by the report has been performed and materials used and installed are in every particular in accordance with and in conformity to the duly approved plans and specifications.

(2) Core Tests. Take core tests of the completed masonry construction in accordance with Section 2-2423(d), Title 24.

The owner's inspector or testing agency is to inspect the coring of the masonry walls and prepare a report of coring operations for the testing laboratory files and mail one copy to the Office of the State Architect. State in this report the number, the location and the condition of all cores cut on the project. Pay particular attention to the description of the bond between the grout fill and the cell walls of the masonry unit. The report should also include a description of any difficulties encountered in the coring operation which might impair the strength of the sample

Submit all cores to the testing laboratory for examination.

If specifically requested by the architect or structural engineer, one third of the cores should be tested for the bond strength of the joint between the masonry units and the grout. This test determines the unit force required to shear the masonry unit face shells from the grout core for each face.

4.9.8 Consolidation of Grout

U.B.C. Sec. 2404 (f) 2.

2. Construction requirements. Reinforcement shall be placed prior to grouting. Bolts shall be accurately set with templates or by approved equivelant means and held in place to prevent movement.

Segregation of the grout materials and damage to the masonry shall be avoided during the grouting process.

Grout shall be consolidated by mechanical vibration during placing before loss of plasticity in a manner to fill the grout space. Grout pours greater than 12 inches shall be reconsolidated by mechanical vibration to minimize voids due to water loss. Grout pours 12 inches or less in height shall be mechanically vibrated, or puddled.

In one-story buildings having wood-frame exterior walls, foundations not over 24 inches high measured from the top of the footing may be constructed of hollow masonry units laid in running bond without mortared head joints. Any standard shape unit may be used, provided the masonry units permit horizontal flow of grout to adjacent units. Grout shall be solidly poured to the full height in one lift and shall be puddled or mechanically vibrated.

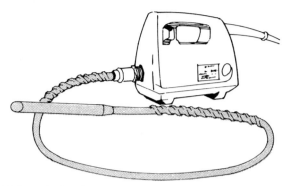

Figure 4.64 *Grout vibrator.*

Grout shall be consolidated by means of a mechanical vibrator if the lift is more than 12 inches. The vibrator, shown in **Figure 4.64**, is usually on a flexible cable with the head from 3/4'' to 1-1/2'' in width. While the vibrator is on it need only be lowered into the grout and slowly removed. If cells are congested with steel, adjacent grouted cells can be consolidated by vibration.

The volume of grout to be consolidated in the cells is relatively small and short bursts of vibration should be adequate.

Consolidation of grout is necessary after excess water is absorbed into the masonry. There is a film of water between the masonry shell and the grout; by compacting the grout it closes up this space causing the grout to have intimate contact with the shell and achieve bonding.

4.9.9   Fluid Mortar for Grout

U.B.C. Sec. 2404 (f) 2.

In nonstructural elements, including fireplaces and residential chimneys, which do not exceed 8 feet in height above the highest point of lateral support, mortar of pouring consistency may be substituted for grout when the masonry is constructed and grouted in pours of 12 inches or less. [See **Figure 4.65.**]

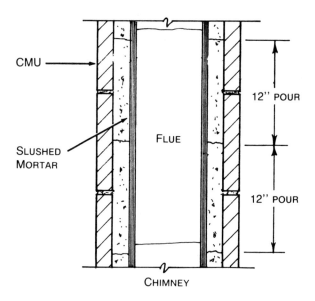

Figure 4.65 *Mortar slushed into non-structural element.*

4.9.10   Grout Barriers

U.B.C. Sec. 2404 (f) 2.

In multiwythe grouted masonry vertical barriers of masonry shall be built across the grout space. The grouting of any section of wall between barriers shall be completed in one day with no interruption longer than one hour. [See Figure 4.66.]

In previous editions of the Uniform Building Code, it was recommended that the spacing of vertical barriers not exceed 30 feet.

4.9.11   Use of Aluminum Equipment

Grout pumped through aluminum pipes will cause an abrasion of the interior of the pipe. This abrasion will cause aluminum particles to

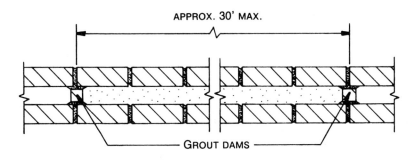

APPROX. 30' MAX.

GROUT DAMS

Figure 4.66 *Grout flow barriers in brick masonry.*

be mixed with the grout and may reduce strength and cause expansion of the grout. Aluminum powder or particles can react with cement and create hydrogen gas, which expands.

U.B.C. Sec. 2404 (g)

(g) **Aluminum Equipment.** Grout shall not be handled nor pumped utilizing aluminum equipment unless it can be demonstrated with the materials and equipment to be used that there will be no deleterious effect on the strength of the grout.

4.9.12 Pumping Grout

Grout is usually placed into masonry walls using a grout pump. The grout is loaded into a grout pump directly from a transit mix truck and then pumped into the masonry cells through a long hose.

Lime or fly ash is sometimes used to aid pumping. The use of fly ash will sometimes save on cement.

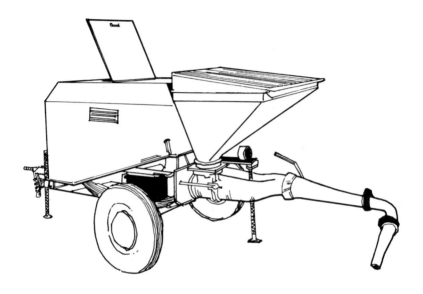

Figure 4.67 *Grout pump.*

Grout pumps are specifically made to pump high slump grout and do not handle concrete, which is stiffer and has larger aggregate. **Figure 4.67** shows a typical grout pump.

4.10 BRACING OF WALLS

Masonry walls are the responsibility of the mason contractor until accepted by the general contractor or owner. Accordingly, bracing masonry walls is up to the contractor and is the general responsibility of either the mason contractor or the general contractor.

It is recommended, and some jurisdictions require, that walls be braced during construction to prevent damage or collapse by wind or other forces.

4.11 PIPES AND CONDUITS EMBEDDED IN MASONRY

U.B.C. Sec. 2407 (g)

(g) Pipes and Conduits Embedded in Masonry. Pipe or conduit shall not be embedded in any masonry so as to reduce the capacity to less than that necessary for required stability or required fire protection.

> **EXCEPTIONS:** 1. Rigid electric conduits may be embedded in structural masonry when their location has been detailed on the approved plan.
>
> 2. Any pipe or conduit may pass vertically or horizontally through any masonry by means of a sleeve at least large enough to pass any hub or coupling on the pipe line. Such sleeves shall be placed not closer than three diameters, center to center, nor shall they unduly impair the strength of construction.
>
> 3. Placement of pipes or conduits in unfilled cores of hollow unit masonry shall not be considered as embedment.

4.12 ADJACENT WORK

Bolts, anchors and other inserts which attach adjoining construction to the walls should be embedded in mortar at the face shell and solidly grouted for the entire remaining embedment in the walls. Where possible, they should be wired to the reinforcing bars to keep them from dislodging during consolidation of the grout.

Roof flashing should penetrate the mortar joints not more than one inch. Metal door frames should be set and braced in-place before the masonry walls are erected. They should be anchored and solidly grouted in-place as the wall is constructed.

4.13 INTERSECTING STRUCTURAL ELEMENTS

4.13.1 Wall to Wall

It is often advantageous for a wall to be designed to work structurally with another intersecting wall or to a roof or floor. The intersecting structural elements must conform to U.B.C. Section 2407 (e).

U.B.C. Sec. 2407 (e) 1.

(e) **Structural Continuity.** 1. **General.** Intersecting structural elements intended to act as a unit shall be anchored together to resist the design forces.

Figure 4.68 shows a few typical details that may be used to tie together corners or intersecting walls.

4.13.2 Walls to Floor or Roof

U.B.C. Sec. 2407 (e) 2.

2. **Wall intersecting with floors or roofs.** Walls shall be anchored to all floors, roofs or other elements which provide lateral support for the wall. Where floors or roofs are designed to transmit horizontal forces to walls, the anchorage to the wall shall be designed to resist the horizontal force. [See **Figure 4.69** and 4.70.]

Precast floor planks are sometimes used on an interior block wall. The planks change directions of span. A concrete topping is cast after the upper wall has been erected. Vertical reinforcing steel goes through the wall and is anchored in the topping concrete.

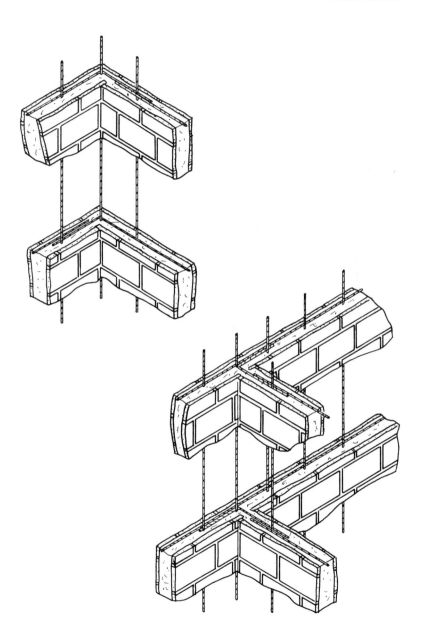

Figure 4.68 Concrete block intersecting wall and corner details. Stagger laps of bars in alternate courses. Lap all bars a minimum of 30 bar diam. or 24″, whichever is greater.

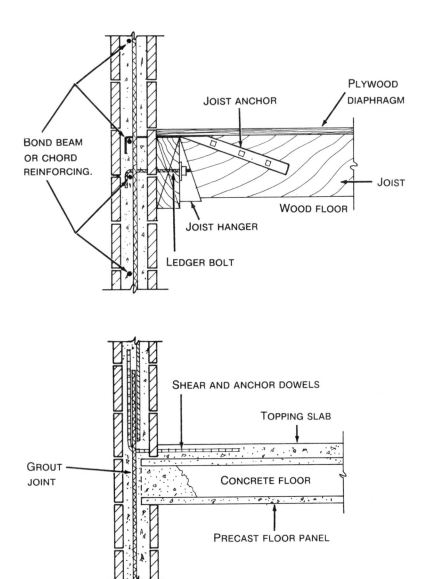

Figure 4.69 *Floor to side wall connection details.*

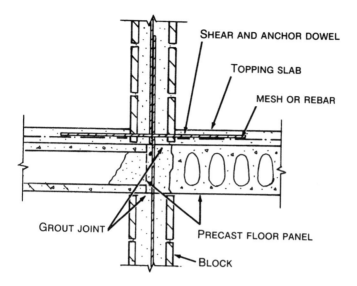

Figure 4.70 *Through floor and wall connection details.*

U.B.C. Sec. 2310.

Anchorage of Concrete or Masonry Walls

Sec. 2310. Concrete or masonry walls shall be anchored to all floors and roofs which provide lateral support for the wall. Such anchorage shall provide a positive direct connection capable of resisting the horizontal forces specified in this chapter or a minimum force of 200 pounds per lineal foot of wall, whichever is greater. Walls shall be designed to resist bending between anchors where the anchor spacing exceeds 4 feet. Required anchors in masonry walls of hollow units or cavity walls shall be embedded in a reinforced grouted structural element of the wall. See Sections 2312 (g), 2312 (h) 2 H. and 2312 (h) 2 I.

4.14 MULTI-WYTHE WALLS

4.14.1 General

U.B.C. Sec. 2407 (e) 3. A.

3. Masonry elements. A. Multiwythe walls. All wythes shall be bonded by grout or tied together by corrosion-resistant metal ties or joint reinforcement conforming to the requirements of Section 2402 and as follows. [See Figure 4.71.]

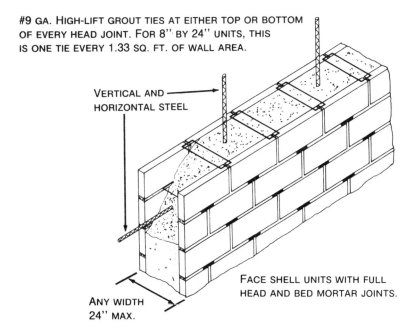

#9 GA. HIGH-LIFT GROUT TIES AT EITHER TOP OR BOTTOM OF EVERY HEAD JOINT. FOR 8" BY 24" UNITS, THIS IS ONE TIE EVERY 1.33 SQ. FT. OF WALL AREA.

VERTICAL AND HORIZONTAL STEEL

FACE SHELL UNITS WITH FULL HEAD AND BED MORTAR JOINTS.

ANY WIDTH 24" MAX.

Figure 4.71 *Multi-wythe component walls.*

U.B.C. Sec. 2404 (f) 2.

In multiwythe grouted masonry, vertical barriers of masonry shall be built across the grout space. The grouting of any section of wall between barriers shall be completed in one day with no interruption longer than one hour. [See **Figure 4.66**.]

4.14.2 Metal Ties for Cavity Wall Construction

U.B.C. Sec. 2407 (e) 3. A. (i).

(i) **Metal ties in cavity wall construction.** Metal ties shall be of sufficient length to engage all wythes. The portion of the tie within the wythe shall be completely embedded in mortar or grout. The ends of the ties shall be bent to 90-degree angles with an extension not less than 2 inches long. Ties not completely embedded in mortar or grout between wythes shall be a single piece of metal with each end engaged in each wythe.

There shall be at least one $\frac{3}{16}$-inch-diameter metal tie for each 4½ square feet of wall area. For cavity walls in which the width of the cavity is greater than 3 inches but not more than 4½ inches, at least one $\frac{3}{16}$-inch-diameter tie for each 3 square feet of wall area shall be provided.

Ties in alternate courses shall be staggered; the maximum vertical distance between ties shall not exceed 24 inches; the maximum horizontal distance between ties shall not exceed 36 inches.

Additional ties spaced not more than 36 inches apart shall be provided around and within 12 inches of the opening.

Metal ties of different size and spacing may be used if they provide equivalent strength between wythes.

4.14.3 Metal Ties for Grouted Multi-Wythe Construction

U.B.C. Sec. 2407 (e) 3. A. (ii)

(ii) **Metal ties for grouted multiwythe construction.** The two wythes shall be bonded together with at least one $\frac{3}{16}$-inch-diameter metal wall tie for each 2 square feet of area. Metal ties of different size and spacing may be used if they provide equivalent strength between wythes. [See Figure 4.72.]

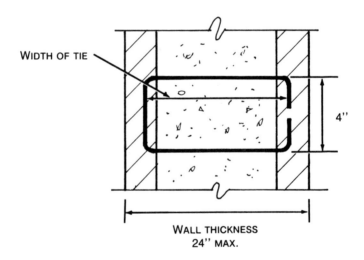

WIDTH OF TIE

4"

WALL THICKNESS
24" MAX.

AREA OF $\frac{3}{16}$" DIAM. WIRES = 0.0276 SQ. IN.
AREA OF 9 GA. WIRES = 0.0346 SQ. IN.

Figure 4.72 Typical positioning of grout tie.

4.14.4   Joint Reinforcing

U.B.C. Sec. 2407 (e) 3 A. (iii)

(iii) **Joint reinforcement.** Prefabricated joint reinforcement for masonry walls shall have at least one crosswire of at least No. 9 gauge for each 2 square feet of wall area. The vertical spacing of the joint reinforcement shall not exceed 16 inches. The longitudinal wires shall be thoroughly embedded in the bed joint mortar. The joint reinforcement shall engage all wythes.

4.14.5   Stack Bond

U.B.C. Sec. 2407 (e) 3. B.

B. **Stack bond.** Where masonry units are laid in stack bond, each wythe shall have longitudinal reinforcement consisting of at least two continuous corrosion-resistant wires with a minimum cross-sectional area of 0.017 square inch each in the horizontal bed joints spaced not more than 16 inches on center vertically. The wires shall be located near the opposite faces of the wythes. [See Figure 4.73.]

4.15   CRACK CONTROL

There are three recommendations that reduce the possibility of having unsightly cracks in concrete masonry walls. These recommendations are:

a. proper jointing

b. proper reinforcing

c. moisture control.

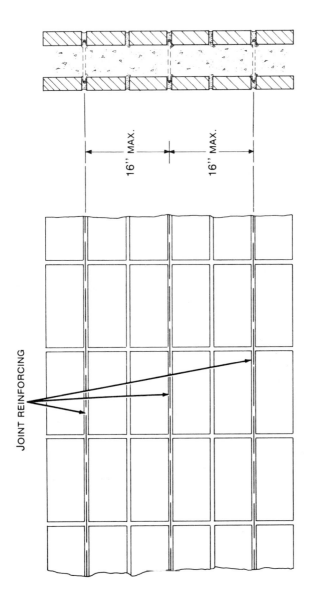

JOINT REINFORCING

16" MAX.

16" MAX.

Figure 4.73 Joint reinforcing in multi-wythe wall laid in stack bond.

194

4.15.1   Jointing; Control Joints and Expansion Joints

All structures are subject to movement and it is vitally necessary to accommodate the possibilities of movement of the building. This movement can occur from a number of sources, such as:

a. temperature changes

b. changes in moisture content or conditions

c. loading conditions

d. foundation movement

e. differential movement of various materials in the building

f. lateral deflections from wind loads

g. earthquake tremors

Masonry walls which are part of the structure are also subject to movement. Masonry walls that move can develop cracking which may be due to the following:

Properties of concrete masonry unit

a. moisture content, i.e. green block at time of laying

b. shrinkage characteristics of the block

c. tensile strength

Environmental factors

a. temperature increases and decreases causing thermal expansion and contraction

Design deficiencies

a. inadequate spacing of horizontal steel or joint reinforcing

b. control joints
   i.   none or too few
   ii.   improperly spaced
   iii. improperly constructed

c. embedded structural steel not properly isolated

4.15.2 Control Joints

Control joints are considered joints that will accommodate shortening, shrinkage and/or reduction in the length of the wall, while expansion joints will accommodate expansion and contraction of the wall or increases and decreases in length.

Control joints are usually vertical and spaced at intervals so that when shortening occurs the resulting cracks will be at the location of the control joints. It is important to locate sufficient control joints so that the movement occurs at the joint rather than through the blocks between the control joints.

Joints in the wall, whether they are control joints or expansion joints, should match any joints that are built into the roof system, the floor system, the spandrel beams, or other elements that are intended to accommodate movement of the building.

When horizontal reinforcing steel is used in the wall, either in bond beams or in the mortar bed with joint reinforcing, the spacing of the control joints to accommodate the shortening of the wall can be increased. Refer to **Table 4-K.**

Table 4-K *Recommended Control Joint Spacing for Moisture Controlled, Type I* Concrete Masonry Units***				
Recommended Spacing of Control Joints	Vertical Spacing of Joint or Horizontal Reinforcement			
	48 in. (1.2 m)	24 in. (0.6 m)	16 in. (0.4 m)	8 in. (0.2 m)
Expressed as Ratio of Panel Length to Height L/H	2	2½	3	4
With Panel Length (L) Not to Exceed:	40 ft. (12 m)	45 ft. (13.7 m)	50 ft. (15 m)	60 ft. (18.3 m)

* The spacing of control joints should be reduced for Type II nonmoisture-controlled units. It may be reduced by one-half or an amount based on experience or practice in the area where the project is located.

**Reference No. 20

Vertical control joints should be located in masonry walls at the following locations:

a. at determined intervals and spacing for the length of the wall

b. at major changes in the wall height

c. at changes in the wall thickness

d. at control joints in the foundation floor and roof

e. at wall openings

f. at wall intersections.

Control joints should be constructed with a vertical head joint, raking back the mortar at least one inch and interrupting the horizontal steel at least at every other bar or joint reinforcing. To prevent the wall from displacing perpendicular to the plane of the wall, dowels may be used across the joint provided one end is encased in a plastic sleeve or pipe. Solid grouted walls crack at the control joint and provide aggregate interlock which prevents displacement and slip, dowels are not required. Primary chord steel, such as bond beams and lintels, must not be cut. The raked vertical head joint should then be caulked to keep it weatherproof.

Typical caulking compounds can stretch best when the width of the joint is greater than the depth of the sealant, similar to a rubber band. Manufacturer's recommendations shall be followed. The usual practice is to place the caulking so the depth of sealant is only half the width. Sealant depth is controlled by using a compatible backup rod. See **Figure 4.74.**

4.15.3 Expansion Joints

Expansion joints are used to accommodate increases in length in long runs of walls and where there are large temperature swings. Spacing of expansion joints should be between 150 and 200 feet and should be located with consideration to the shape and plan of the structure. Concern must be given to the expansion of the wall and the possibility of pushing out the end walls, thus, the expansion joints should be near the ends of the building. Many conditions allow the expansion joints to be in the middle between ends of the walls, allowing movement of the walls both ways towards the center.

CONTROL JOINT IN A WALL

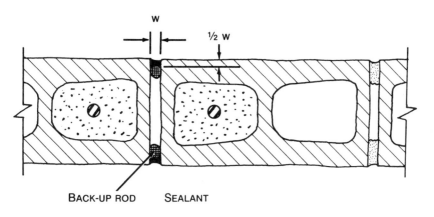

BACK-UP ROD   SEALANT

CONTROL JOINT AT WALL ADJOINING
A PERPINDICULAR WALL.

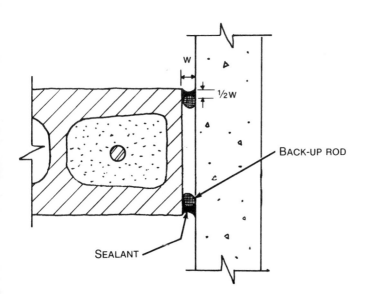

BACK-UP ROD

SEALANT

Figure 4.74 Control joints in walls.

The expansion joint should be filled with a caulking material that will both expand and compress and allow for total movement of the wall.

Expansion joints are similar to control joints in their appearance. However, a control joint provides for movement of a wall in contraction and can have some solid masonry or grout in the joint. An expansion joint is used to provide for movement of a wall as it expands and should not have any incompressible material in it.

4.15.4 Summary

Crack control for concrete masonry walls is as follows:

1. Use a moisture controlled unit in climatic balance. This means that the moisture condition of the concrete masonry units is in a state of equilibrium with the relative humidity of the project site.

2. Use joint reinforcing. This is effective by being located in the face shells.

3. Put in adequate properly-spaced and properly-constructed control joints and expansion joints.

The knowledge of the mason contractor and the masonry inspector can be used to alert the general contractor and the architect concerning potential problem areas for crack control. If potential problems are pointed out ahead of time, the responsibility of the mason contractor is reduced.

4.15.5 Crack Repair

If a crack does occur, the repair depends on where it is, what kind of block is involved, and how important appearance is. Usually it is not possible to accomplish a repair with only paint or a clear sealer. Repeated movement will reopen cracks that have been bridged over. The crack must be opened up enough so that a bead of flexible caulking can be applied. The caulking should remain flexible even when dry. It can be painted over with a compatible paint. Some caulkings come in a range of colors and can be matched to mortar colors when used on integral color block jobs.

4.16 COLD WEATHER MASONRY CONSTRUCTION

4.16.1 General

The following cold weather provisions were prepared by the International Masonry Industry All Weather Council and are based on "Recommended Practice Guide Specification for Cold Weather Masonry, 1970."

U.B.C. Sec. 2404 (c). 1.

(c) **Cold Weather Construction. 1. General.** All materials shall be delivered in a usable condition and stored to prevent wetting by capillary action, rain and snow.

The tops of all walls not enclosed or sheltered shall be covered with a strong weather-resistive material at the end of each day or shutdown.

Partially completed walls shall be covered at all times when work is not in progress. Covers shall be draped over the wall and extend a minimum of 2 feet down both sides and shall be securely held in place, except when additional protection is required in Section 2404 (c) 4.

4.16.2 Preparation

U.B.C. Sec. 2404 (c) 2.

2. **Execution—Preparation.** If ice or snow has inadvertently formed on masonry bed, it shall be thawed by application of heat carefully applied until top surface of the masonry is dry to the touch.

A section of masonry deemed frozen and damaged shall be removed before continuing construction of that section.

4.16.3 Construction

> **U.B.C. Sec. 2404 (c) 3.**
>
> **3. Construction.** Masonry units shall be dry. Wet or frozen masonry units shall not be laid.
>
> Air temperature 40°F. to 32°F.: Sand or mixing water shall be heated to produce mortar temperatures between 40°F. and 120°F.
>
> Air temperature 32°F. to 25°F.: Sand and mixing water shall be heated to produce mortar temperatures between 40°F. and 120°F. Maintain temperatures of mortar on boards above freezing.
>
> Air temperature 25°F. to 20°F.: Sand and mixing water shall be heated to produce mortar temperatures between 40°F. and 120°F. Maintain mortar temperatures on boards above freezing. Salamanders or other sources of heat shall be used on both sides of walls under construction. Windbreaks shall be employed when wind is in excess of 15 mph.
>
> Air temperature 20°F. and below: Sand and mixing water shall be heated to provide mortar temperatures between 40°F. and 120°F. Enclosure and auxiliary heat shall be provided to maintain air temperature above 32°F. Temperature of units when laid shall be not less than 20°F.

4.16.4 Protection

> **U.B.C. Sec. 2404 (c) 4.**
>
> **4. Protection.** When the mean daily air temperature is 40°F. to 32°F., masonry shall be protected from rain or snow for 24 hours by covering with weather-resistive membrane.

When the mean daily air temperature is 25°F. to 20°F., masonry shall be completely covered with insulating blankets or equally protected for 24 hours.

When the mean daily air temperatue is 20°F. and below, masonry shall be maintained above 32°F. for 24 hours by enclosure and supplementary heat, by electric heating blankets, infrared heat lamps or other approved methods.

4.16.5 Placing Grout and Protection of Grouted Masonry

U.B.C. Sec. 2404 (c) 5.

5. Placing grout and protection of grouted masonry. When air temperatures fall below 40°F., grout mixing water and aggregate shall be heated to produce grout temperatures between 40°F. and 120°F.

Masonry to be grouted shall be maintained above freezing during grout placement and for at least 24 hours after placement.

Where atmospheric temperatures fall below 20°F., enclosures shall be provided around the masonry during grout placement and for at least 24 hours after placement.

4.16.6 Summary of Recommended Cold Weather Practices*

The following points are important factors in laying masonry work in cold weather.

- Planning and scheduling of the work must be done beforehand if masonry work is to be built in cold temperatures.

- Take advantage of warm days by working on the outside of the structure, saving the inside work for the colder days.

---

*Ref. No. 10

- Store all masonry units close to the structure. Be sure units are covered and off the ground to prevent moisture or frost from penetrating.

- Build a mortar mixing area covered with a roof. Keep the sand pile covered to prevent moisture, ice or snow from penetrating.

- Preheat the sand and water before mixing the mortar.

- Use Type III, high early strength portland cement or a non-chloride accelerator additive for a quicker set.

- Do not use antifreeze in the mortar. Calcium chloride is considered to be an accelerator not an antifreeze. Never use calcium chloride if there is metal in the mortar joints.

- Preheat the masonry units before laying them if the temperature is very cold or if frost or ice is present.

- Build protective shelters such as windbreaks and enclosures to protect the mason and the masonry work.

- Observe good safety practices when building shelters to prevent them from collapsing or blowing over, causing damage and injury.

- Take protective measures at the end of the workday to protect the work and to ensure that work is started on time the next day. Some protective measures are covering the work and piles of materials, draining the hoses and cleaning out the mortar pans. Placing a block of wood in the water barrel so the water does not freeze keeps the barrel from deforming.

4.17 HOT WEATHER MASONRY CONSTRUCTION*

4.17.1 General

Building with masonry in hot weather (90° F. and over) can cause special problems. High temperatures can cause the materials to become very warm, affecting their performance. Rapid evaporation will

*Ref. No. 9

also occur that will have an effect on hydration and curing. Special consideration must be given to the handling and selection of materials and to construction procedures during hot weather.

4.17.2 Performance

The physical properties of masonry will change with an increase in temperature:

1. Bond strength can be poor as units are hotter and drier, causing an increase in suction rate.

2. The compressive strength of the mortar and grout will be weaker as water is quickly evaporated, leaving little for hydration.

3. Workability is affected, as more water is required in the mortar for constant consistency and in grout to make filling of spaces possible.

4. Heat will affect the amount of air entraining used, since more is required in hot weather.

5. The initial and final sets of mortar will occur faster.

6. Water will evaporate quickly on a joint's exterior surfaces, causing a decrease in strength and durability.

7. The initial water content of mortar will be higher, but the placing will be difficult and the time period will be short.

4.17.3 Handling and Selection of Materials

When hot weather is expected, the materials should be stored in a shaded or cool place. Increasing the cement content will cause the mortar and grout to gain strength quickly. The amount of lime can be increased giving the mortar a higher water retentivity.

Covering the aggregate with a light color or clear plastic sheet will retard the evaporation of water. Adding extra water will help keep the aggregate cool as evaporation has a cooling effect.

The units used should be stored in the shade and covered. The use of cold water on the units, especially block, for cooling purposes is not recommended. The use of cold water or ice water as mixing water will lower the temperature of the mortar.

4.17.4 Construction Procedure

When placing masonry units in hot weather, special consideration should be given to all equipment that comes in contact with the mortar. Flushing the mixers, tools and mortar boards occasionally with cold water helps keep temperature to a minimum.

Mortar should not be mixed too far ahead, and when mixed, should be stored in a cool, shady place. Avoid placing long mortar beds ahead of the units as bond is reduced. When extremely high temperatures are expected, consideration should be given to stopping placement of masonry during the hottest times of day.

4.18 WET WEATHER MASONRY CONSTRUCTION*

4.18.1 General

Building with masonry in rainy weather is possible if some type of shelter or covering is provided.

4.18.2 Performance

Rain can cause excessive wetting of materials, affecting their performance. The change in unit moisture content can cause considerable change in dimension. The amount will vary with the type of material used. Moisture will also reduce the absorptive quality of the units so that poor bond occurs between the units and the mortar. Water will evaporate more slowly so less mixing water need be added.

---

*Ref. No. 9

If it rains on a building before the mortar is set, the cementitious material can be washed out, reducing strength and resulting in the possible collapse of the joint. The mortar may also be washed over the faces of the units causing a staining effect.

4.18.3   Construction Procedures

Building can be done in wet weather providing rain does not fall on the masonry materials or on the freshly laid walls. The cement, units and sand should be covered to keep them dry. They should also be stored off the ground so there is no migration of moisture from the ground to the materials.

A masonry wall, built in rainy conditions, should be built under or behind a shelter. This can be in the form of a roof or floor slab, or inside an enclosure similar to the type used in cold weather. Walls should be protected from rain for 24 to 48 hours, depending on the temperature, so that the mortar is fully set and bond has occurred.

4.18.4   Protection of Masonry*

Partially completed masonry walls that are exposed to rain may become so saturated with water that they require months to dry out. This can cause efflorescence or water soaking into the framing.

While the masonry walls are being built, it is the responsibility of the mason to be sure that the walls are covered at all times (when not being worked on). The covering can be of plastic, canvas, or some other suitable material that not only covers the top of the wall but hangs over at least two feet on the face. It should also be weighted down to prevent the wind from getting under it and damaging the wall. The common practice of laying a heavy board on top of the wall at the end of the workday does not keep the work protected and can cause the masonry underneath to sag or bow out of position.

---

*Ref. No. 11

4.19 REINFORCED CONCRETE MASONRY INSPECTION CHECKLIST

MATERIAL

☐ 1. Check the following points against the plans:

　☐ (a) Are deviations allowed by the plan check engineer?

　☐ (b) Is continuous inspection necessary?

　☐ (c) Are called inspections necessary?

☐ 2. Check the type and quality of CMU used.

☐ 3. Check concrete masonry units for

　☐ (a) correct size and type, (per U.B.C. Standard Nos. 24–4, 24–6),

　☐ (b) curing (per U.B.C. Standard Nos. 24–4, 24–6),

　☐ (c) cleanliness,

　☐ (d) soundness (per U.B.C. Standard Nos. 24–4, 24–6).

☐ 4. Check quality of the units by test specimens and find whether

　☐ (a) a laboratory test is required,

　☐ (b) required inspection holes are provided.

☐ 5. Make sure that the CMU conforms to the U.B.C. Standards (U.B.C. Standard Nos. 24–4, 24–6).

☐ 6. Check to see that the structure complies with the plans and check for

　☐ (a) strength of the masonry,

　☐ (b) stress.

☐   7. Check for separation between building.

☐   8. Check thickness of the walls.

☐   9. Check the size of bond beams.

☐ 10. Check the reinforcing steel for

       ☐ (a) kind and grade,

       ☐ (b) size (U.B.C. No. 2407(h)4., U.B.C. No. 2409(e)1),

       ☐ (c) location (U.B.C. No. 2407(h)3.b.),

       ☐ (d) bracing,

       ☐ (e) clearances (U.B.C. No. 2407(e)2.),

       ☐ (f) deformation,

       ☐ (g) additional steel around openings (U.B.C. No. 2407(h)3.B.),

       ☐ (h) placed within allowable tolerances (U.B.C. No. 2404(e)).

☐ 11. Check the following points for connections

       ☐ (a) size and location of joist anchors,

       ☐ (b) size, location and number of bolts,

       ☐ (c) size and location of dowels,

       ☐ (d) location of stirrups,

       ☐ (e) veneer ties (if any).

☐ 12. Check the placing of headers and lintels of material other than masonry.

WORKMANSHIP

☐ 13. See that material is properly stored off the ground and covered with waterproof material.

☐ 14. Check to see that work is kept dry at all times.

☐ 15. Check to see that clean water is used.

☐ 16. Classify the mortar by type and use (U.B.C. Table No. 24–A).

☐ 17. Verify proportions of the mortar mix and time of mixing.

☐ 18. Check sand for

  ☐ (a) cleanliness,

  ☐ (b) quality and fineness,

  ☐ (c) compliance with code requirements (U.B.C. Standard No. 24–21).

☐ 19. Verify that cement meets requirements of the U.B.C. Standards (U.B.C. Standard No. 24–20).

☐ 20. Verify that aggregates meet the requirements of U.B.C. Standards (U.B.C. Standard No. 24–21).

☐ 21. Check lime for conformance to the U.B.C. Standards (U.B.C. Standard No. 24–17).

☐ 22. Make certain that water is clean and free from harmful substances.

☐ 23. Check plasticizing agents for conformance to Standards.

☐ 24. Verify that admixtures conform to the following requirements

  ☐ (a) have been approved,

  ☐ (b) are of right quantity,

  ☐ (c) are not used with plastic cement.

☐ 25. Sample panels have been provided and approved, if required.

☐ 26. Check the consistency of mortar.

☐ 27. Verify that mortar is properly handled in mixing.

☐ 28. See that mortar is not excessively retempered.

☐ 29. Check grout for

    ☐ (a) proportions (U.B.C. Table No. 24–B),

    ☐ (b) consistency,

    ☐ (c) compressive strength (U.B.C. Standard No. 24–29),

    ☐ (d) handling,

    ☐ (e) segregation.

☐ 30. Verify that all head, bed and wall joints are

    ☐ (a) watertight,

    ☐ (b) correct size,

    ☐ (c) solidly filled with mortar,

    ☐ (d) buttered where required.

☐ 31. Joint size and type of joint are as required.

☐ 32. In checking head, bed or end joints, verify that

    ☐ (a) head or bed joints are of proper size,

    ☐ (b) head joints are completely filled (Exception: U.B.C. No. 2404(d)4.),

    ☐ (c) end joints are filled with mortar.

☐ 33. Check joints where fresh masonry is joined to set masonry.

☐ 34. Check reinforced hollow unit masonry for

    ☐ (a) vertical alignment and continuity of cells,

    ☐ (b) lapping (U.B.C. No's. 2409(e)3a., 2409(e)6),

    ☐ (c) leakage of grout,

    ☐ (d) cleanout openings for pours over 5' (U.B.C. No. 2404(f)1., U.B.C. Table No. 24–B),

    ☐ (e) overhanging mortar,

    ☐ (f) sealing of cleanout cells,

    ☐ (g) position of reinforcement,

    ☐ (h) requirements when work is stopped for one hour or longer.

CONSTRUCTION

☐ 35. Check unprotected steel supporting members for

    ☐ (a) correct location of mechanical installation supports,

    ☐ (b) size and location of bolts, rivets and connections,

    ☐ (c) size and spacing of bracing connections,

    ☐ (d) size and alignment of connection holes,

    ☐ (e) shims and dry packing,

    ☐ (f) location and size of stiffeners,

    ☐ (g) size and alignment of base plates.

☐ 36. Check the following points in inspection of bearing on solid masonry (U.B.C. No. 2407(c)1.A.).

    ☐ (a) suitability of bearing masonry,

    ☐ (b) size of bearing masonry,

    ☐ (c) size, length, placement and embedment of connectors,

    ☐ (d) location of bolt ties.

☐ 37. Verify proper sill material and check for anchorage of supporting members to footings.

☐ 38. Verify the following points in checking anchoring of wood floor joists to supporting masonry members.

    ☐ (a) required size of ledges,

    ☐ (b) required size, spacing and length of bolts and joist anchors.

☐ 39. Where floor joists are parallel to the wall, check for

    ☐ (a) placing of required blocking,

    ☐ (b) type of anchors required,

    ☐ (c) use of proper connections to the anchors.

☐ 40. Verify the following in inspecting floor joists entering a masonry wall.

    ☐ (a) required size, spacing and bearing of joists,

    ☐ (b) required air space around joists,

    ☐ (c) required bridging and/or blocking,

    ☐ (d) embedment of joist in wall,

    ☐ (e) required connectors for anchors,

    ☐ (f) required anchors.

☐ 41. Verify the following points in the inspection of a masonry building where fire-resistive floors are required.

    ☐ (a) proper material for fire resistance,

    ☐ (b) required thickness of floor slab,

    ☐ (c) required supports,

    ☐ (d) required reinforcing,

    ☐ (e) required length of time forms remain in place for concrete floors.

☐ 42. Contraction joints and control joints are located and provided as indicated or required.

☐ 43. Weepholes are provided if required.

☐ 44. Cover the following points in checking for height of a masonry building.

    ☐ (a) approved material,

    ☐ (b) placement and spacing of steel,

    ☐ (c) correct grouting procedure.

☐ 45. Verify the following when inspecting chases,

    ☐ (a) location and spacing on approved plans,

    ☐ (b) purpose,

    ☐ (c) maximum permitted depth,

    ☐ (d) no reduction of the required strength and fire resistance of the wall.

☐ 46. Check bearing walls for

    ☐ (a) end support,

    ☐ (b) footings,

    ☐ (c) thickness of walls,

    ☐ (d) bond beams,

    ☐ (e) materials used,

    ☐ (f) joints,

    ☐ (g) reinforcing,

    ☐ (h) grout,

    ☐ (i) shoring (when required),

    ☐ (j) anchorage.

☐ 47. Inspect non-bearing walls for

    ☐ (a) location, height and thickness of walls,

    ☐ (b) approved materials,

    ☐ (c) type of units used,

    ☐ (d) proper placement of ties, anchors and bolts.

☐ 48. Where there is a change of thickness in non-bearing walls, be sure to

    ☐ (a) locate the position on the approved plans,

    ☐ (b) check for compliance of required top plates,

    ☐ (c) verify location of ties, anchors, bolts and blocking.

☐ 49. Check racking and toothing at wall intersections.

☐ 50. Check wall ties.

☐ 51. Check corners and returns.

☐ 52. Check masonry on concrete for

    ☐ (a) width and depth of footing excavations,

    ☐ (b) type and grade of masonry units,

    ☐ (c) grouting and metal inserts,

    ☐ (d) anchorage around main steel,

    ☐ (e) embedment of ties or connection to main steel,

    ☐ (f) type, spacing and material of ties.

☐ 53. Check corbeling for

    ☐ (a) maximum projections,

    ☐ (b) bonding and anchorage,

    ☐ (c) required temporary supports,

    ☐ (d) required reinforcing.

☐ 54. Pointing, replacement of defective units, and repair of other defects are promptly performed.

☐ 55. Waterproofing of walls is performed as required.

☐ 56. Methods of final cleaning are as required.

Masonry Units

5.1 ICBO EVALUATION SERVICE, INC. EVALUATION REPORTS

Table 5-A shows evaluation reports issued by ICBO ES for masonry materials, systems and admixtures. It is necessary to check the current status of evaluation reports to determine if they are still active and up-to-date.

5.2 TYPICAL CONCRETE MASONRY UNITS

5.2.1 Precision Units

Given in this section are typical concrete masonry units that are used in construction. They are manufactured in accordance with U.B.C. Standards Nos. 24-4 and 24-6, and are illustrated in **Figures 5.1** through **5.8**.

5.2.2 Slumped Block

Slumped units, which have the appearance of Old California or Spanish adobe, are available in standard and special sizes and in a variety of colors. Slumped block dimensions may vary by more than 1/2'' from the specified standard dimensions. The variations in dimensions are adjusted by different size mortar joints to obtain the nominal sizes shown. Refer to **Figure 5.9** and **Table 5-B**.

Table 5-A   ICBO Evaluation Reports

COMPANY	NO.	PRODUCT	Date of Last Action
ITW Ramset/Red Head Michigan City, IN	1372	Phillips Red Head Sleeve Anchors	June 88
Tomax Corporation Phoenix, AZ	PFC-2864	Tomax Masonry Panels	Sept. 88
W.R. Grace & Co.	1041	Vermiculite Concrete and Masonry Walls	June 88
Dr. Juan Haener San Diego, CA	2996	Haener Mortarless Concrete Blocks	Aug. 83
CMACN Citrus Heights, CA	2848P	Standard Details for 1-story reinforced CMU Residential Construction	Mar. 86
CMACN Citrus Heights, CA	4115	Strength Design of 1 to 4-story CMU Buildings	Sept. 87
Graystone Block, Inc. Modesto, CA	3606	Solar Blok and Solar Stone Hollow Load-Bearing CMU	Dec. 87
Korfil, Incorporated West Brookfield, MA	3773	Korfil I and II Insulated Hollow CMU	July 88
Addiment Incorporated Atlanta, GA	4271	Mortarplast AR 13 Mortar Additive	Feb. 88

Table 5-A ICBO Evaluation Reports

COMPANY	NO.	PRODUCT	Date of Last Action
American Colloid Company Arlington Heights, IL	3759	Easy/Spred Plasticizer for Mortar	Nov. 87
Fiber Mesh Inc. Chattanooga, TN	NER-284	Fibermesh Fiber	Feb. 88
North American Mortar Supply, Inc. Edmonton, Alberta, Canada	4199	EX PL 84-6 Mortar Retardar	Nov. 86
Morton Chemical, Inc. Chicago, IL	3462	Mor-Ad 336, 366 and 434 Adhesives	Mar. 87
Molly Fasteners, Emhart Corp. Temple, PA	2350	Parabolt Concrete Anchors	Nov. 86
CalMat Company Los Angeles, CA	3294	Colton Mortar Cement	Aug. 85
ITW Ramset Rolling Meadows, IL	2391	Trubolt Wedge, Auk, Dynabolt, Dynaset and Ram-Drill Anchors.	Aug. 88
Ultra Block Southwest, Inc. Tucson, AZ	4523	Ultra Block Masonry Units	Jan. 88

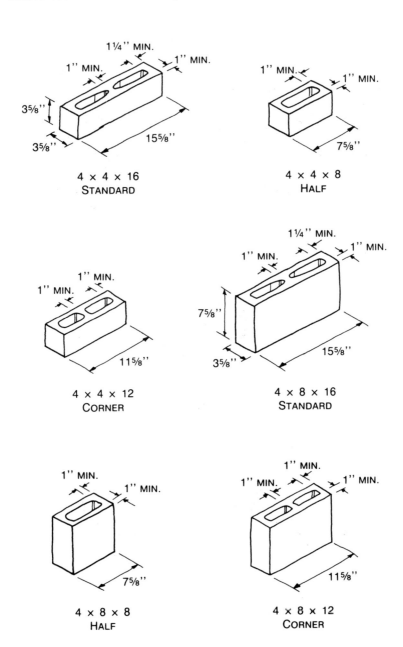

Figure 5.1 *Four-inch-wide concrete masonry units.*

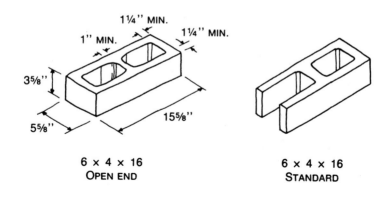

6 × 4 × 16
OPEN END

6 × 4 × 16
STANDARD

6 × 4 × 8
HALF

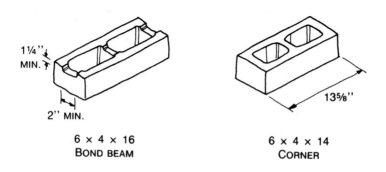

6 × 4 × 16
BOND BEAM

6 × 4 × 14
CORNER

Figure 5.2 Six-inch-wide concrete masonry units.

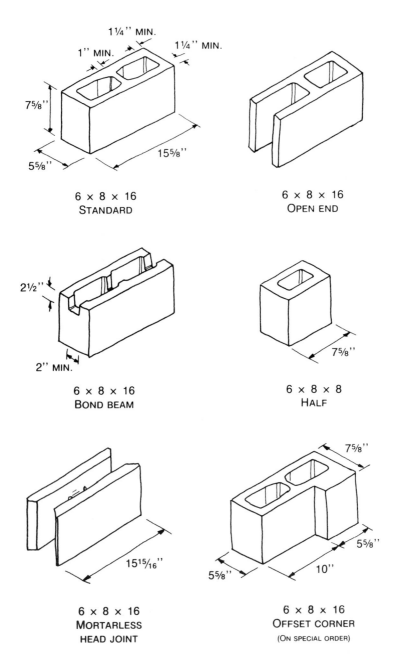

Figure 5.2 *(Continued) Six-inch-wide concrete masonry units.*

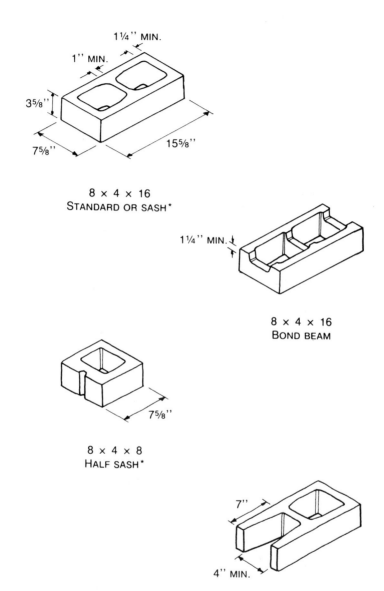

1¼" MIN.

1" MIN.

3⅝"

7⅝"

15⅝"

8 × 4 × 16
STANDARD OR SASH*

1¼" MIN.

8 × 4 × 16
BOND BEAM

7⅝"

8 × 4 × 8
HALF SASH*

7"

4" MIN.

*DIMENSIONS FOR SASH SLOT VARY
AMONG PRODUCERS. SEE YOUR
MANUFACTURER FOR SPECIFICATIONS.

8 × 4 × 16
OPEN END

Figure 5.3 Eight-inch-wide concrete masonry units.

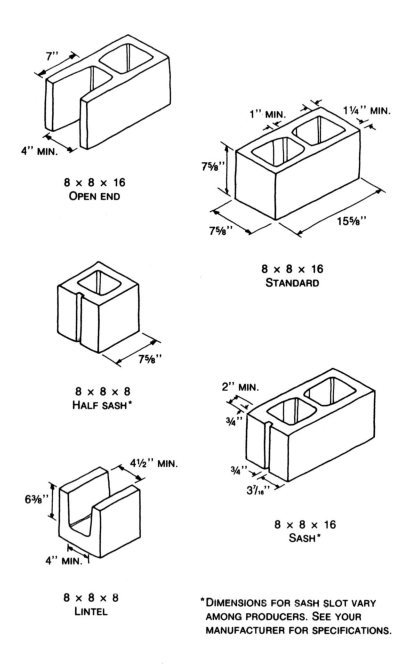

7"

4" MIN.

8 × 8 × 16
OPEN END

1" MIN.   1¼" MIN.

7⅝"

7⅝"   15⅝"

8 × 8 × 16
STANDARD

7⅝"

8 × 8 × 8
HALF SASH*

2" MIN.

¾"

¾"

3⁷/₁₆"

8 × 8 × 16
SASH*

4½" MIN.

6⅜"

4" MIN.

8 × 8 × 8
LINTEL

*DIMENSIONS FOR SASH SLOT VARY
AMONG PRODUCERS. SEE YOUR
MANUFACTURER FOR SPECIFICATIONS.

Figure 5.3 *(Continued) Eight-inch-wide concrete masonry units.*

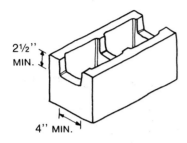

2½" MIN.

4" MIN.

8 × 8 × 16
BOND BEAM

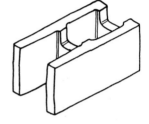

8 × 8 × 16
OPEN END
BOND BEAM

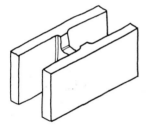

8 × 8 × 16
DOUBLE OPEN END
BOND BEAM

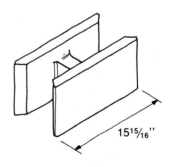

15¹⁵⁄₁₆"

8 × 8 × 16
MORTARLESS
HEAD JOINT

Figure 5.3 (Continued)

225

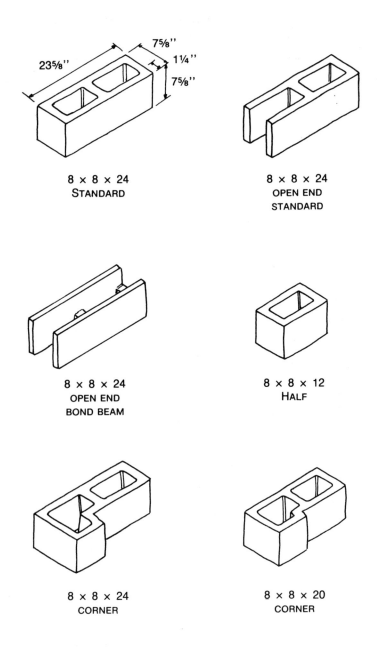

Figure 5.4 *Eight-inch-wide by 24" long concrete masonry units.*

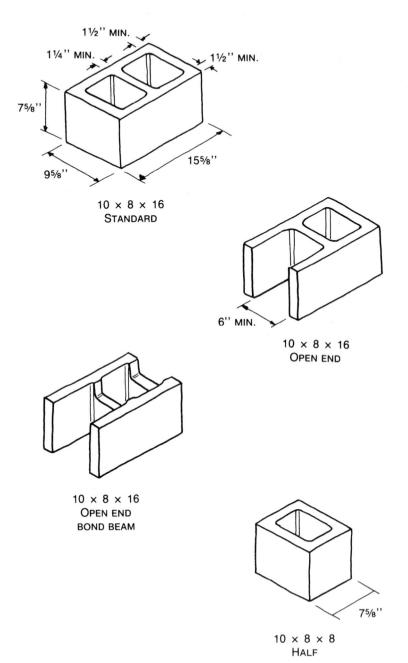

1½'' MIN.

1¼'' MIN.

1½'' MIN.

7⅝''

9⅝''

15⅝''

10 × 8 × 16
STANDARD

6'' MIN.

10 × 8 × 16
OPEN END

10 × 8 × 16
OPEN END
BOND BEAM

7⅝''

10 × 8 × 8
HALF

Figure 5.5 *Ten-inch CMU.*

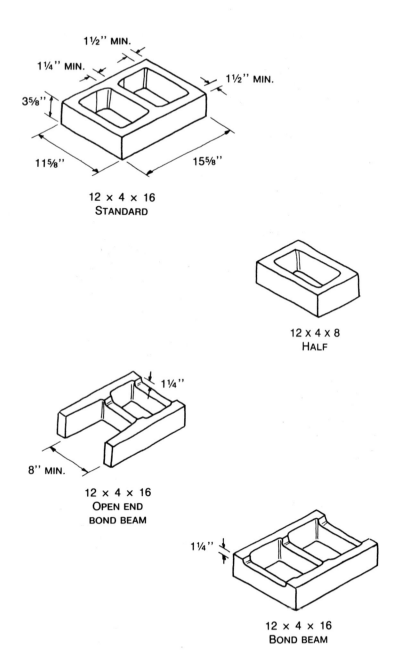

1½" MIN.

1¼" MIN.

1½" MIN.

3⅝"

11⅝"

15⅝"

12 x 4 x 16
STANDARD

12 X 4 X 8
HALF

1¼"

8" MIN.

12 x 4 x 16
OPEN END
BOND BEAM

1¼"

12 x 4 x 16
BOND BEAM

Figure 5.6 *Twelve-inch CMU, four inches high.*

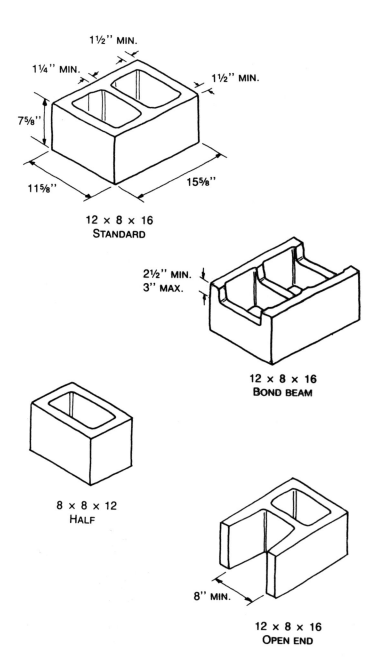

1½" MIN.

1¼" MIN.

1½" MIN.

7⅝"

11⅝"

15⅝"

12 × 8 × 16
STANDARD

2½" MIN.
3" MAX.

12 × 8 × 16
BOND BEAM

8 × 8 × 12
HALF

8" MIN.

12 × 8 × 16
OPEN END

Figure 5.7 Twelve-inch CMU, eight inches high.

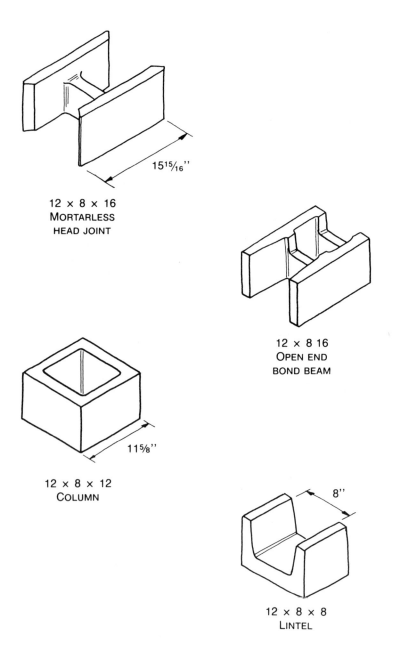

12 × 8 × 16
MORTARLESS
HEAD JOINT

12 × 8 16
OPEN END
BOND BEAM

12 × 8 × 12
COLUMN

12 × 8 × 8
LINTEL

Figure 5.7 *(Continued) Twelve-inch CMU, eight inches high.*

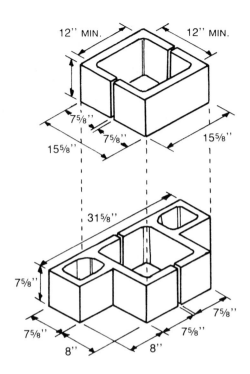

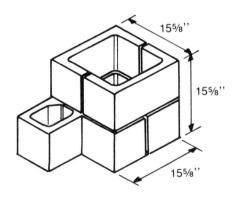

Figure 5.8 Pilaster block. Pilasters, or "column" block units, are readily available. The two most common pilasters used for 8" wide walls are the 12" x 16" and the 16" x 16" (illustrated).

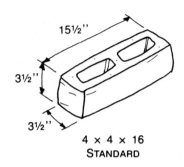

15½"

3½"

3½"

4 × 4 × 16
STANDARD

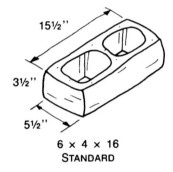

15½"

3½"

5½"

6 × 4 × 16
STANDARD

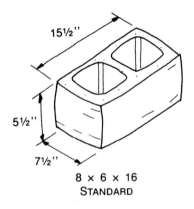

15½"

5½"

7½"

8 × 6 × 16
STANDARD

Figure 5.9
Slumped block.

Table 5-B *Slumped Block Nominal Dimensions (width, height, length)*

Dimensions	Description
4 × 4 × 16	solid
4 × 4 × 12	solid corner
4 × 4 × 8	solid half
4 × 6 × 16	solid
6 × 4 × 16	standard and bond beam
6 × 4 × 14	corner
6 × 4 × 12	three quarter
6 × 4 × 8	half
6 × 6 × 16	standard and bond beam
6 × 6 × 8	half
8 × 4 × 16	standard, bond beam, open end, open end bond beam
8 × 4 × 8	half
8 × 6 × 16	standard, bond beam, open end, open end bond beam
8 × 6 × 8	half
12 × 4 × 16	standard, open end, open end bond beam
12 × 4 × 12	corner/column
12 × 4 × 8	half
12 × 6 × 16	standard and bond beam
12 × 6 × 8	half
16 × 4 × 16	column
16 × 6 × 16	column
12 × 6 × 12	column
6 × 2 × 16	cap
8 × 2 × 16	cap
6 × 4 × 16	cap
8 × 4 × 16	cap

5.2.3 Custom Face Units

The custom face units shown in **Figure 5.10** are a small sampling of the broad range of concrete masonry architectural units available from the industry on special order.

5.2.4 Split Face Units

Split face block is manufactured as a unit that is normally made double and is literally split apart on a splitter; a machine which resembles a guillotine. The splitter has blades at the top and bottom (and sometimes at the sides) which exert pressure on the blocks, breaking them apart.

Many factors determine the look of the split face, both as to size variances and the amount of aggregate exposure. Split face block is intended to have a rougher texture than precision block. Various configurations of block such as flutted, scored, etc., will split in a different manner than a full split face. The vertical perpendicularity of scored and fluted split face block is subject to variation.

The split face units shown in **Figure 5.11** are a small sampling of the broad range of concrete masonry architectural units available from the industry on special order. Depths and widths of scores vary. Consult a local manufacturer for specific information.

5.2.5 Special Proprietary Units

Special concrete masonry units have been created and used when special requirements must be met. These requirements can be greater sound control, more energy efficiency, or increased ease of placement.

Sound blocks are made with slotted openings in one side of the face shells. This allows sound waves to enter the cell where the sound waves reflect back and forth. Desirable acoustical ratings are then achieved, especially in high noise areas such as large furnace rooms or gymnasiums (see **Figure 2.3**).

Energy efficiency can also be improved when special concrete masonry units are used that contain insulation inside the face shells,

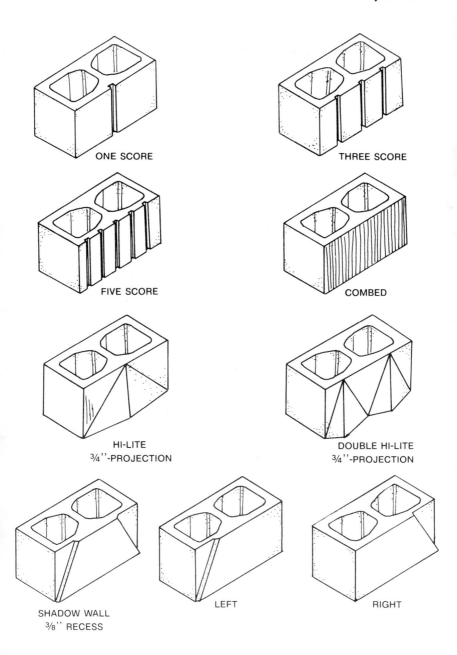

ONE SCORE

THREE SCORE

FIVE SCORE

COMBED

HI-LITE
¾''-PROJECTION

DOUBLE HI-LITE
¾''-PROJECTION

SHADOW WALL
⅜'' RECESS

LEFT

RIGHT

Figure 5.10 Custom face units.

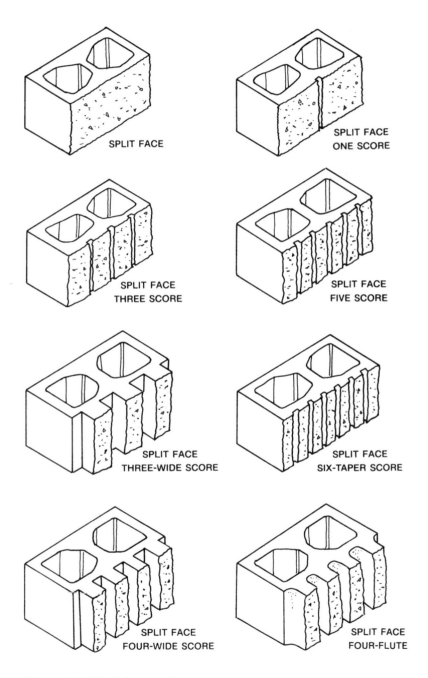

Figure 5.11 Split face units.

usually as part of the unit or as inserts. The insulation helps reduce unwanted transfer of energy from one side of the wall to the other (see **Figure** 2.3).

Mortarless units have also been developed. These block are an attempt to reduce the cost of masonry construction. The block units either interlock without mortar or use thin high-bond mortars, sometimes only on the bed joints. Mortarless units are usually designed to be grouted (see **Figure** 2.3).

5.3   LENGTH, HEIGHT AND QUANTITIES IN CONCRETE MASONRY WALLS

5.3.1   Length and Height of Walls

Table 5-C shows the number of blocks of various sizes in walls of certain length and height. For **Table 5-C** to be interpreted properly, the following information should be noted.

- For exact wall length subtract thickness of one mortar joint.

- For exact opening dimensions add thickness of one mortar joint to height and width.

- When using combinations of 8'' high and 4'' high blocks, a detailed wall section should be made to establish height dimensions.

- The modular chart shown is for 4'' and 8'' high units. Six inch high precision and slump block units are available generally throughout the market. Further freedom in dimensioning is now possible using 6'' high units to start or story out in combination with the above chart.

---

*Ref. No. 7

Table 5-C   Number of Blocks in a Wall
⅜" Horizontal and Vertical Mortar Joints

LENGTH	NO. 16" LONG BLOCKS	HEIGHT	NO. 4" HIGH BLOCKS	NO. 8" HIGH BLOCKS
0'-8''	½	0'-4''	1	
1'-4''	1	0'-8''	2	1
2'-0''	1½	1'-0''	3	
2'-8''	2	1'-4''	4	2
3'-4''	2½	1'-8''	5	
4'-0''	3	2'-0''	6	3
4'-8''	3½	2'-4''	7	
5'-4''	4	2'-8''	8	4
6'-0''	4½	3'-0''	9	
6'-8''	5	3'-4''	10	5
7'-4''	5½	3'-8''	11	
8'-0''	6	4'-0''	12	6
8'-8''	6½	4'-4''	13	
9'-4''	7	4'-8''	14	7
10'-0''	7½	5'-0''	15	
10'-8''	8	5'-4''	16	8
11'-4''	8½	5'-8''	17	
12'-0''	9	6'-0''	18	9
12'-8''	9½	6'-4''	19	
13'-4''	10	6'-8''	20	10
14'-0''	10½	7'-0''	21	
14'-8''	11	7'-4''	22	11
15'-4''	11½	7'-8''	23	
16'-0''	12	8'-0''	24	12
16'-8''	12½	8'-4''	25	
17'-4''	13	8'-8''	26	13
18'-0''	13½	9'-0''	27	
18'-8''	14	9'-4''	28	14
19'-4''	14½	9'-8''	29	
20'-0''	15	10'-0''	30	15
20'-8''	15½	10'-4''	31	
21'-4''	16	10'-8''	32	16

Table 5-C Number of Blocks in a Wall
(Cont.)   ⅜" Horizontal and Vertical Mortar Joints

LENGTH	NO. 16" LONG BLOCKS	HEIGHT	NO. 4" HIGH BLOCKS	NO. 8" HIGH BLOCKS
22'-0"	16½	11'-0"	33	
22'-8"	17	11'-4"	34	17
23'-4"	17½	11'-8"	35	
24'-0"	18	12'-0"	36	18
24'-8"	18½	12'-4"	37	
25'-4"	19	12'-8"	38	19
26'-0"	19½	13'-0"	39	
26'-8"	20	13'-4"	40	20
27'-4"	20½	13'-8"	41	
28'-0"	21	14'-0"	42	21
28'-8"	21½	14'-4"	43	
29'-4"	22	14'-8"	44	22
30'-0"	22½	15'-0"	45	
30'-8"	23	15'-4"	46	23
31'-4"	23½	15'-8"	47	
32'-0"	24	16'-0"	48	24
32'-8"	24½	16'-4"	49	
40'-0"	30	16'-8"	50	25
50'-0"	37½	17'-0"	51	
60'-0"	45	17'-4"	52	26
70'-0"	52½	17'-8"	53	
80'-0"	60	18'-0"	54	27
90'-0"	67½	18'-4"	55	
100'-0"	75	18'-8"	56	28
200'-0"	150	19'-0"	57	
300'-0"	225	19'-4"	58	29
400'-0"	300	19'-8"	59	
500'-0"	375	20'-0"	60	30

5.3.2 Quantities of Materials

The following list gives examples of the quantities of different materials required for various jobs:

- When figuring concrete-sand-gravel mixtures, figure one cubic yard of material for 80 sq. feet by 3-1/2'' thick.

- Five cubic feet of mortar will lay approximately 100 8'' x 8'' x 16'' concrete blocks.

- One cubic foot of mortar will lay approximately 100 bricks.

- One cubic yard of stucco will cover approximately 320 sq. feet surface 1'' thick.

- One ton of wall rock will cover approximately 50 sq. feet.

- Six bricks are required per square foot in a 4'' thick wall. In a patio or walk, it takes 4-1/2 bricks per sq. ft. of surface.

- Slate weighs approximately 5 lbs. per square foot.

- Flagstone covers about 1 square foot with each 15 lbs.

- When mixing mortar, mix 1 part portland cement, 1/2 part hydrated lime to 4-1/2 parts sand. (Type S mortar)

Table 5-D gives grout requirements for concrete block construction.

Table 5-D   Grout Quantities

Standard Two Cell Block*	Grouted Cells Vert. Steel Spacing	Cu. Yds. ▲ of Grout Per 100 Sq. Ft. of Wall	Cu. Yds. per ▲ 100 Block (8'' High) (16'' Long)	Block per Cu. Yd. (8'' High) (16'' Long)
6'' THICK WALLS	All Cells Filled	0.93	0.83	120
	16'' O.C.	0.55	0.49	205
	24'' O.C.	0.42	0.37	270
	32'' O.C.	0.35	0.31	320
	40'' O.C.	0.31	0.28	360
	48'' O.C.	0.28	0.25	396
8'' THICK WALLS	All Cells Filled	1.12	1.00	100
	16'' O.C.	0.65	0.58	171
	24'' O.C.	0.50	0.44	225
	32'' O.C.	0.43	0.38	267
	40'' O.C.	0.37	0.33	300
	48'' O.C.	0.28	0.30	330
10'' THICK WALLS	All Cells Filled	1.38	1.23	80
	16'' O.C.	0.82	0.73	137
	24'' O.C.	0.63	0.56	180
	32'' O.C.	0.53	0.47	214
	40'' O.C.	0.47	0.42	240
	48'' O.C.	0.43	0.38	264
12'' THICK WALLS	All Cells Filled	1.73	1.54	65
	16'' O.C.	1.01	0.90	111
	24'' O.C.	0.76	0.68	146
	32'' O.C.	0.64	0.57	174
	40'' O.C.	0.57	0.51	195
	48'' O.C.	0.53	0.47	215

* For open end block add 10% more grout.
For slumped block deduct 5% grout.
Horizontal bond beams assumed spaced 4' O.C.

▲ A 3% allowance has been included for loss and job conditions.

Glossary of Terms

ABSORPTION — The amount of water the unit will absorb when immersed in either cold or boiling water for a stated length of time. It is expressed in pounds of water per cubic foot of net volume for concrete blocks.

ADMIXTURES — Materials added to cement, aggregate and water such as water-repellants, air-entraining or plasticizing aids, pigments, or aids to retard or speed up setting.

AGGREGATE — Inert particles such as sand, gravel and rock, which when bound together with portland cement and water form concrete.

AMERICAN SOCIETY FOR TESTING AND MATERIALS (ASTM) — A voluntary organization that sets standards for testing and materials based on a concensus agreement.

ANCHOR TIES — Any type of fastener used to secure masonry to some stable object, such as another wall, usually for tension value.

AREAS:

Bedded Area — The area of the surface of a masonry unit which is in contact with mortar in the plane of the joint.

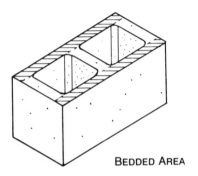

BEDDED AREA

Effective Area of Reinforcement (A_s) — The cross-sectional area of reinforcement multiplied by the cosine of the angle between the reinforcement and the direction for which effective area is to be determined.

Gross Cross-Sectional Area — The total area of a section perpendicular to the direction of the load, including areas within cells and within re-entrant spaces unless these spaces are to be occupied in the masonry by portions of adjacent masonry. (Note: The gross cross-sectional area of scored units is determined to the outside of the scoring.)

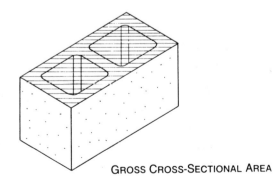

GROSS CROSS-SECTIONAL AREA

Net Cross-Sectional Area — The gross cross-sectional area of a section minus the average area of ungrouted cores or cellular spaces. (Note: the cross-sectional area of grooves in scored units is not usually deducted from the gross cross-sectional area to obtain the net cross-sectional area.)

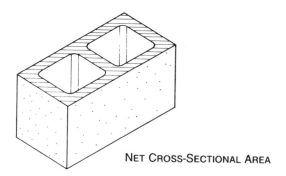

NET CROSS-SECTIONAL AREA

Transformed Area — The equivalent area of one material to a second based on the ratio of moduli of elasticity of the first material to the second.

BAT — A piece used in the wall smaller than the length of the field units being used. Generally used to fill out an odd dimension, and can be either cut with a saw or brick hammer.

BEARING WALL — Any masonry wall which supports more than 200 pounds per lineal foot superimposed load, or any such wall supporting its own weight for more than one story.

BOND:

Adhesion Bond — The adhesion between masonry units and mortar or grout.

Common Bond — Units laid so that they lap half over each other in successive courses. Also called half bond.

Mechanical Bond — Units laid so that they lap over each other in successive courses. Includes quarter bond, third bond, and half or common bond.

Reinforcing Bond — The adhesion between steel reinforcement and mortar or grout.

Running Bond — Lapping of units in successive courses so that the vertical head joints lap. Placing vertical mortar joints centered over the unit below is called center bond, or half bond, while lapping ⅓ or ¼ is called third bond or quarter bond.

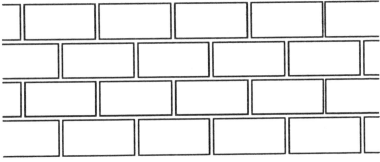

RUNNING BOND

Stack Bond — A bonding pattern where no unit overlaps either the one above or below; all head joints form a continuous vertical line. Also called plumb joint bond, straight stack, jack bond, jack on jack, and checkerboard bond.

STACK BOND

BOND BEAM — One or more courses of masonry units poured solid and reinforced with longitudinal reinforcing bars that are designed to take the stress resulting from lateral forces.

BUTTERING — Spreading mortar on masonry units with a trowel.

CAP — Masonry units laid on top of finished masonry wall or pier. Metal caps are units formed of metal, as for flashing.

CAVITY WALL — A wall built of two or more wythes of masonry units so arranged as to provide a continuous air space within the wall. The facing and backing, outer wythes, are tied together with non-corrosive ties, e.g. brick or wire.

CELL (CORE) — The molded open space in a concrete masonry unit having a gross cross-sectional area greater than 1½ square inches.

CHASE — A continuous recess built into a wall to receive pipes, ducts, etc.

CLEANOUT — An inspection hole at the base of a cell used to clean out debris and inspect steel placement. Minimum size shall measure not less than 2'' × 3''.

CLOSER — The last unit laid in a course. A closer may be a whole unit or one that is shorter and usually appears in the field of the wall.

COLLAR JOINT — The vertical, longitudinal joint between two wythes of masonry.

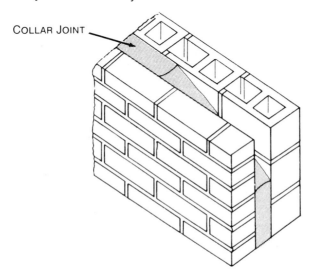

COLLAR JOINT

COLUMN — A vertical structural member whose horizontal dimension measured at right angles to the thickness does not exceed three times the thickness.

COMPOSITE WALL — Reinforced grouted masonry wall in which inner and outer wythes are dissimilar materials, i.e. block and brick, block and glazed structural units, etc.

COMPRESSIVE STRENGTH — The maximum load required to fracture the masonry unit by applying a compressive force to the upper and lower surface of the unit. Expressed as either gross compressive strength or net compressive strength.

CONCRETE MASONRY UNIT:

A-Block — A hollow unit with one end closed and the opposite end open, forming two cells when laid in the wall; also called open end block.

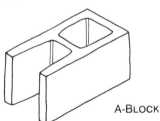

A-BLOCK

Bond Beam Block — A hollow unit with portions depressed 2¼'' or more to permit the forming of a continuous channel for horizontal reinforcing steel and grout.

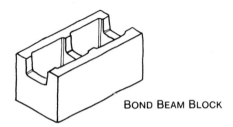

BOND BEAM BLOCK

Channel Block — A hollow unit with portions depressed less than 1¼'' to permit the forming of a continuous channel for reinforcing steel and grout.

H-Block — A hollow unit with both ends open commonly called a double open end.

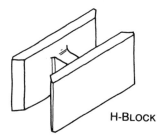

H-BLOCK

Offset Block — A unit which is not rectangular in shape.

Open-End Block — A term applied to both H-blocks and A-blocks. In the figure shown, it is commonly called open end unit.

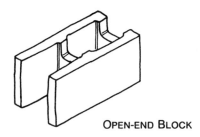

OPEN-END BLOCK

Pilaster Block — Concrete masonry units designed for use in construction of plain or reinforced concrete masonry pilasters and columns.

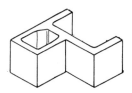

OPEN CENTER PILASTER
OR BANJO BLOCK

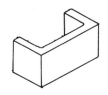

PILASTER ALTERNATE
OR C BLOCK

PILASTER BLOCKS

Return L Block — Concrete masonry unit designed for use in corner construction for various thickness walls.

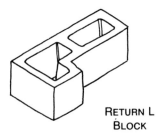

RETURN L
BLOCK

Sash Block — A concrete masonry unit which has an end slot for use in openings to receive metal window frames and pre-molded expansion joint material.

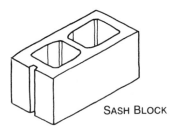

SASH BLOCK

Scored Block — Block with grooves to provide patterns, as for example to simulate raked joints, available in architectural face units.

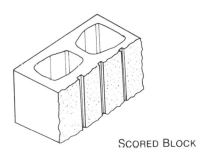

SCORED BLOCK

Sculptured Block — Block with specially formed surfaces, in the manner of sculpturing.

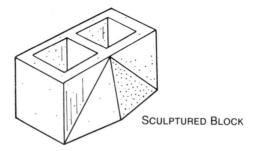

SCULPTURED BLOCK

Shadow Block — Block with face formed in planes to develop surface patterns.

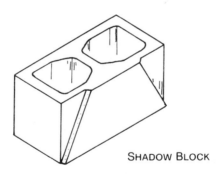

SHADOW BLOCK

Sill Block — A solid concrete masonry unit used for sills or openings.

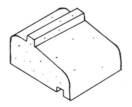

SILL BLOCK

Slump Block — Concrete masonry units produced so they "slump" or sag irregular fashion before they harden.

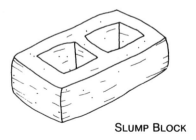

SLUMP BLOCK

Solid Unit — Masonry units in which the vertical cores are less than 25% of the cross-sectional area.

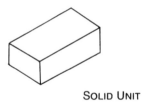

SOLID UNIT

Split Face Block — Concrete masonry units with one or more faces having a fractured surface for use in masonry wall construction.

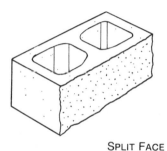

SPLIT FACE

COPING — The material or units used to form a cap or finish on top of a wall, pier or pilaster.

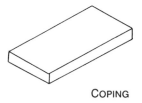

COPING

CORBEL — A shelf or ledge formed by projecting successive courses of masonry out from the face of the wall.

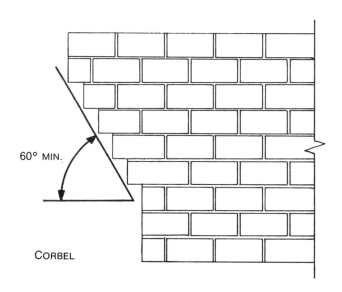

60° MIN.

CORBEL

CORE — See **CELL**.

COURSE — A continuous horizontal layer of masonry units.

CURING — The maintenance of proper conditions of moisture and temperature during initial set to develop required strength and reduce shrinkage in concrete products.

DIMENSIONS:

Actual Dimensions — The measured dimensions of a designated item, for example, a designated masonry unit or wall as used in the structure. The actual dimension shall not vary from the specified dimension by more than the amount allowed in the appropriate standard in Section 2402 of the 1988 U.B.C.

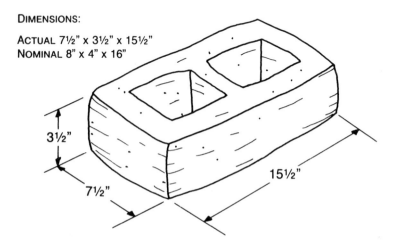

DIMENSIONS:

ACTUAL 7½" x 3½" x 15½"
NOMINAL 8" x 4" x 16"

3½"

7½"

15½"

Nominal Dimensions — Generally equal to its specified dimension plus the thickness of the joint with which the unit is to be laid.

Specified Dimensions — The dimensions specified for the manufacture or construction of masonry, masonry units, joints, or any other component of a structure. Unless otherwise stated, all calculations shall be made using or based on specified dimensions.

EFFLORESCENCE — A whitish powder resulting from the leaching of soluble sulfate salts on the surface of masonry.

FACE SHELL — The side wall of a hollow concrete masonry unit.

FACED WALL — A wall in which the facing and the backing are so bonded or otherwise tied as to act as a composite element.

FACING — Any material forming a load-bearing part of a wall and used as a finish surface (veneer takes no load other than its own weight).

FASCIA — The flat outside horizontal member of a cornice.

FAT MORTAR — See **MORTAR**.

FIRE CLAY — A finely-ground clay.

FIRE WALL — Any wall that sub-divides a building so as to resist the spread of fire.

FIRE-RESISTIVE — In the absence of a specific ruling by the authority having jurisdiction, the term "fire-resistive" is applied to all building materials which are not combustible in temperatures of ordinary fires and will withstand such without serious impairment of their usefulness for at least one hour, maybe two hours, three hours, four hours, etc.

FLOW AFTER SUCTION — Flow of mortar measured after subjecting it to a vacuum produced by a head of two inches of mercury. The suction apparatus and its use is described in Sections 27 and 28 of ASTM C91.

FLOW MORTAR — Measure of mortar consistency sometimes termed the initial flow determined on the flow table described in ASTM C230. Use of the flow table and method of calculating the flow is described in Section 9 of ASTM C109.

f'$_m$ — The specified compressive strength of a masonry prism at the age of 28 days.

FURRING — A method of finishing the face of a masonry wall to provide space for insulation, prevent moisture transmittance, or to provide a surface for finishing.

GROSS COMPRESSIVE STRENGTH — The compressive strength of the unit based on the total area as defined in "Cross-Sectional Area." Expressed in pounds per square inch (psi).

GROUT — A mixture of cement, aggregate, water, and sometimes an admixture, which is poured into bond beams and vertical cells to encase the steel and to bond units together.

GROUT LIFT — An increment of grout height within the total pour; a pour may consist of one or more grout lifts.

GROUT POUR — The total height of a masonry wall to be poured prior to the erection of additional masonry. A grout pour will consist of one or more grout lifts.

HOLLOW MASONRY UNIT — A masonry unit whose net cross-sectional area in any plane parallel to the bearing surface is less than 75% of its cross-sectional area measured in the same plane.

IMPERVIOUSNESS — The quality of resisting moisture penetration.

JOINT REINFORCEMENT — Steel wire, bar or prefabricated reinforcement which is placed in mortar bed joint.

JOINTING — The process of finishing mortar joints with a tool.

JOINTS:

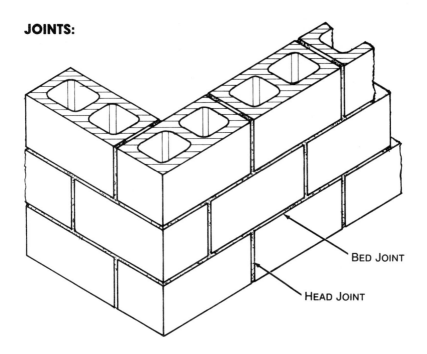

Bed Joint

Head Joint

Bed Joint — The mortar joint that is horizontal at the time the masonry units are placed.

Collar Joint — The vertical space separating a wythe of masonry from another wythe or from another continuous material and may be filled with mortar or grout.

Control Joint — A continuous unbonded masonry joint to control the location and amount of separation resulting from the contraction of the masonry wall so as to avoid the development of excessively high stresses and cracking in the masonry.

Dry Joint — Head or bed joint without mortar.

Expansion Joint — A vertical joint or space to allow for movement due to volume changes.

Head Joint — The mortar joint between units in the same wythe, usually vertical, sometimes called the cross joint.

Shrinkage Joint — See **CONTROL JOINT.**

LEAD — The section of a wall built up and racked back on successive courses at a corner or end of a wall. The line is attached to the leads and the wall is then built up between them.

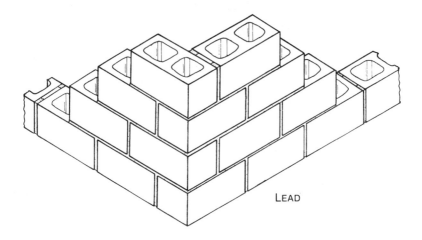

LEAD

LEAN MORTAR — See **MORTAR**.

LIFT — The increment of grout height within the total pour. A pour may consist of one or more grout lifts.

LIME:

Hydrated Lime — Quicklime treated with only enough water to satisfy its chemical demand. Packaged in a powdered form, does not require slaking.

Lime Putty — Slaked quicklime.

Processed Lime — Pulverized quicklime, which must be slaked and cooled prior to use.

Quicklime — A hot or unslaked lime.

LINTEL — A beam placed over an opening in a wall.

LOAD-BEARING WALL — Any wall which in addition to supporting its own weight supports the building above it.

MASONRY UNIT — Brick, tile, stone, glass block or concrete block conforming to the requirements specified in Section 2402.

Hollow Masonry Unit — A masonry unit whose net cross-sectional area in every plane parallel to the bearing surface is less tan 75% of the gross cross-sectional area in the same plane.

Solid Masonry Unit — A masonry unit whose net cross-sectional area in every plane parallel to the bearing surface is 75% or more of the gross cross-sectional area in the same plane.

MITER — A cut made at an angle, two members to fit together at an angle.

MODULAR DIMENSION — A dimension based on a given module, usually 8 inches in the case of concrete block masonry.

MODULAR MASONRY UNIT — A masonry unit whose actual dimensions are one mortar joint less than the modular dimension, i.e.: 8" × 8" × 16" is actually 7⅝" × 7⅝" × 15⅝", to allow for ⅜" joints.

MORTAR — A plastic mixture of cementitious materials, fine aggregate and water, with or without the inclusion of other specified materials.

Fat Mortar — A mortar that tends to be sticky and adheres to the trowel.

Harsh Mortar — A mortar that, due to an insufficiency of plasticizing material, is difficult to spread.

Lean Mortar — A mortar that, due to a deficiency of cementious material, is harsh and difficult to spread.

NET CROSS-SECTIONAL AREA — The gross cross-sectioned area of a section minus the average area of ungrouted cores or cellular spaces. (Note: the cross-sectional area of grooves in scored units is not usually deducted from the gross cross-sectional area to obtain the net cross-sectional area.)

NOMINAL DIMENSION — A dimension which may vary from the actual dimension by the thickness of a mortar joint but not more than ½ inch. (¾''–⅞'' for some slump units.) The actual dimension is usually ⅜ inch less than nominal in concrete masonry units.

PARAPET — The part of a wall that extends above the roof level.

PARGING — The process of applying a coat of cement mortar to the back of the facing material or the face of the backing material; sometimes referred to as pargeting.

PERMEABILITY — The quality of allowing the passage of fluids.

PILASTER — An integral portion of the wall which projects on one or both sides and acts as a vertical beam, a column, an architectural feature, or any combination thereof.

PLUMB JOINT BOND — See **STACK BOND**.

POINTING — Filing mortar into a joint after the masonry unit is laid.

PRISM — Units mortared together, generally in stack bond, forming a wallette or assemblage to simulate ''in wall construction,'' grouted or ungrouted per specification requirements. This is the standard test sample for determination of f'_m.

QUICKLIME — See **LIME**.

RACKING — A method of building the end of a wall by stepping back each course so that it can be built on to and against without toothing; also used in corner leads.

RACKING

REBAR — Reinforcing steel bars of various sizes and shapes used to strengthen masonry.

REINFORCED HOLLOW CONCRETE MASONRY — Masonry in which reinforcement is embedded in either mortar or grout.

RETEMPER — See **TEMPER**.

ROWLOCK — A brick laid on its face edge so that the normal end of brick is visible in the wall face. Frequently spelled rolok. Sometimes called bull-headers.

RUNNING BOND — See **BOND**.

SHELL — The outer portion of a hollow masonry unit as placed in masonry.

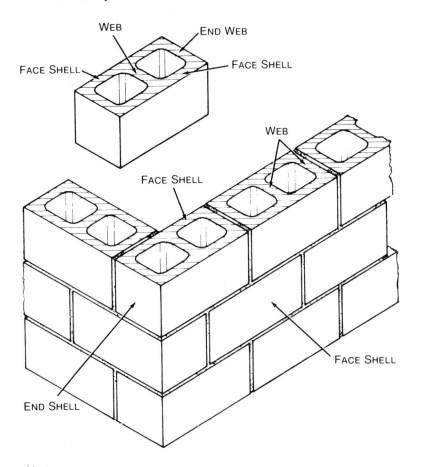

SHOVED JOINT — Head or vertical joints filled by buttering the ends of the units with mortar and shoving them against the units previously laid.

SLAKING — Adding water to quicklime and changing it into lime putty or slaked lime. The addition of water causes the quicklime to heat from hydration and break down into finer pieces.

SLUSHED JOINTS — Head joints filled after units are laid by throwing mortar in with edge of trowel. (Not recommended practice.)

SOFFIT — The underside of a beam, lintel or reveal.

SOLDIER — The underside of a beam, lintel or reveal.

SPANDREL — That part of a wall between the head of a window and the sill of the window above it.

STACK BOND — See **BOND**.

STORY POLE — A marked pole used to establish vertical heights during the construction of the wall.

STRETCHER — A unit laid with its length horizontal and parallel with the face of a wall.

STRINGING MORTAR — The procedure of spreading enough mortar on the bed joint to lay several masonry units.

STRINGING
MORTAR

STRUCK JOINT — In masonry, a mortar joint which is formed with a recess at the bottom of the joint.

TEMPER — To moisten mortar and re-mix to the proper consistency for use. Also called re-tempering.

TOOLING — Compressing and shaping the face of a mortar joint with a special tool other than a trowel. Also called jointing.

TOOTHING — The temporary ending of a wall wherein the units in alternate courses project (usually not permitted).

TUCK POINTING — The filling in with fresh mortar of cut-out or defective mortar joints in masonry.

VENEER — A masonry facing which is attached to the backup but not so bonded as to act with it under load. As opposed to faced wall.

WALLS:

Bonded Wall — A wall in which two or more of its wythes of masonry are adequately bonded together to act as a structural unit.

Cavity Wall — A wall containing continuous air space with a minimum width of 2 inches and a maximum width of 4½ inches between wythes and the wythes are tied together with metal ties.

Hollow-Unit Masonry Wall — That type of construction made with hollow masonry units in which the units are laid and set in mortar.

WALL TIE — A mechanical fastener which connects wythes of masonry to each other or to other materials.

WATER RETENTIVITY — The property of mortar which prevents the rapid loss of water.

WEB — The cross wall connecting the face shells of a hollow concrete masonry unit.

WYTHE — The portion of a wall which is one masonry unit in thickness; also called *tier*. A collar joint is not considered a wythe.

Uniform Building Code Standards

UNIFORM BUILDING CODE STANDARD NO. 24-3
CONCRETE BUILDING BRICK

Based on Standard Specification C 55 (1980)
of the American Society for Testing and Materials

See Section 2402 (b) 5 of the Uniform Building Code

Scope

Sec. 24.301. This standard covers concrete building brick and similar solid units made from portland cement, water and suitable mineral aggregates with or without the inclusion of other materials.

NOTE 1: Concrete brick covered by this standard are made from lightweight or normal weight aggregates, or both.

NOTE 2: When particular features are desired, such as weight classification, high compressive strength, surface textures for appearance or bond, finish, color, fire resistance, insulation, acoustical properties, or other specific features, such properties should be specified separately by the purchaser. However, local sellers should be consulted as to the availability of concrete brick having the desired features.

Classification

Sec. 24.302. (a) **Types.** Two types of concrete brick in each of two grades are covered, as follows:

1. **Type 1, moisture-controlled units.** Concrete brick designated as Type I (Grades N-I and and S-1) shall conform to all requirements of this standard, including the requirements of Table No. 24–3–A.

2. **Type II, nonmoisture-controlled units.** Concrete brick designated as Type II (Grades N–II and S–II) shall conform to all requirements of this standard except the requirement of Table No. 24–3–A.

(b) **Grades.** Concrete brick manufactured in accordance with this standard shall conform to two grades as follows:

1. **Grade N.** For use as architectural veneer and facing units in exterior walls and for use where high strength and resistance to moisture penetration and severe frost action and moisture penetration are desired.

2. **Grade S.** For general use where moderate strength and resistance to frost action and moisture penetration are required.

Materials

Sec. 24.303. (a) **Cementitious Materials.** Materials shall conform to the following applicable U.B.C. Standards:

1. Portland Cements—U.B.C. Standard No. 26-1 modified as follows:

Limitation on insoluble residue—1.5 percent

Limitation on air content of mortar,

Volume percent—22 percent maximum.

Limitation on loss on ignition—7 percent maximum.

Limestone with a minimum 85 percent calcium carbonate (Ca Co$_3$) content may be added to the cement, provided the requirements of U.B.C. Standard No. 26-1 as modified above are met.

2. Blended Cements—U.B.C. Standard No. 26-1.

3. Hydrated Lime, Type S—U.B.C. Standard No. 24-18.

4. Pozzolans—U.B.C. Standard No. 26-9.

(b) **Aggregates.** Aggregates shall conform to the following U.B.C. Standards, except that grading requirements shall not necessarily apply:

1. Normal weight—U.B.C. Standard No. 26-2.

2. Lightweight—U.B.C. Standard No. 26-3.

(c) **Other Constituents.** Air-entraining agents, coloring pigments, integral water repellents, finely ground silica, etc., shall be previously established as suitable for use in concrete and either shall conform to U.B.C. Standard No. 26-9 where applicable, or shall be shown by test or experience not to be detrimental to the durability of the concrete.

Physical Requirements

Sec. 24.304. At the time of delivery to the work site the concrete brick shall conform to the physical requirements prescribed in Table No. 24-3-B.

The moisture content of Type I concrete brick at the time of delivery shall conform to the requirements prescribed in Table No. 24-3-A.

Dimensions and Permissible Variations

Sec. 24.305. Overall dimensions (width, height, or length) shall not differ by more than ⅛ inch from the specified standard dimensions.

NOTE 3: Standard dimensions of concrete brick are the manufacturer's designated dimensions. Nominal dimensions of modular size concrete brick are equal to the standard dimensions plus ⅜ inch, the thickness of one standard mortar joint. Nominal dimensions of nonmodular size concrete brick usually exceed the standard dimensions by ⅛ to ¼ inch.

Variations in thickness of architectural units such as split-faced or slumped units will usually vary from the specified tolerances.

Visual Inspection

Sec. 24.306. (a) **General.** All concrete brick shall be sound and free of cracks or other defects that would interfere with the proper placing of the unit or impair the strength or permanence of the construction. Minor cracks incidental to the usual method of manufacture, or minor chipping resulting from customary methods of handling in shipment and delivery, shall not be deemed grounds for rejection.

(b) **Brick in Exposed Walls.** Where concrete brick are to be used in exposed wall construction, the face or faces that are to be exposed shall be free of chips, cracks or other imperfections, when viewed from 20 feet, except that if not more than 5 percent of a shipment contains slight cracks or small chips not larger than ½ inch, this shall not be deemed grounds for rejection.

Methods of Sampling and Testing

Sec. 24.307. The purchaser or his authorized representative shall be accorded proper facilities to inspect and sample the concrete brick at the place of manufacture from the lots ready for delivery. At least 10 days should be allowed for completion of the test.

Sample and test concrete brick in accordance with U.B.C. Standard No. 24-7, Sampling and Testing Concrete Masonry Units.

The moisture content requirements of Type I units shall be determined in accordance with U.B.C. Standard No. 24-27, Test for Drying Shrinkage of Concrete Block, conducted not more than 12 months prior to delivery of the units.

TABLE NO. 24–3–A—MOISTURE CONTENT REQUIREMENTS FOR TYPE I CONCRETE BRICK

	MOISTURE CONTENT, MAX. PERCENT OF TOTAL ABSORPTION (Average of 3 Concrete Brick)		
	Humidity[1] Conditions at Jobsite or Point of Use		
LINEAR SHRINKAGE, PERCENT	Humid	Intermediate	Arid
0.03 or less	45	40	35
From 0.03 to 0.045	40	35	30
0.045 to 0.065, max.	35	30	25

[1]Arid—Average annual relative humidity less than 50 percent.
Intermediate—Average annual relative humidity 50–75 percent.
Humid—Average annual relative humidity above 75 percent.

TABLE NO. 24–3–B—STRENGTH AND ABSORPTION REQUIREMENTS

COMPRESSIVE STRENGTH, MIN., psi (Concrete Brick Tested Flatwise)			WATER ABSORPTION, MAX, (Avg. of 3 Brick) WITH OVENDRY WEIGHT OF CONCRETE Lb./Ft.³		
Average Gross Area			Weight Classification		
Grade	Avg. of 3 Concrete Brick	Individual Concrete Brick	Lightwieght Less than 105	Medium Weight Less than 125 to 105	Normal Weight 125 or more
N–I	3500	3000	15	13	10
N–II	3500	3000	15	13	10
S–I	2500	2000	18	15	13
S–II	2500	2000	18	15	13

UNIFORM BUILDING CODE STANDARD NO. 24-4

HOLLOW AND SOLID LOAD-BEARING CONCRETE MASONRY UNITS

Based on Standard Specifications C 90–86 of the American Society for Testing and Materials

Scope

Sec. 24.401. This standard covers solid (units with 75 percent or more net area) and hollow load-bearing concrete masonry units made from portland cement, water and mineral aggregates with or without the inclusion of other materials. There are three weight classifications for concrete masonry units: (1) normal weight, (2) medium weight, and (3) lightweight. There are two categories of concrete masonry units: (1) precision, and (2) particular feature units.

NOTE 1: Concrete masonry units covered by this standard are made from lightweight or normal weight aggregates, or both.

NOTE 2: When particular features are desired, such as surface textures, finish, color, special shapes, or when other features are desired such as weight classification, high compressive strength, fire resistance, insulation or acoustical properties, such properties should be specified separately. However, local suppliers should be consulted as to the availability of units having the desired features.

Classification

Sec. 24.402. (a) **Grades.** Concrete masonry units manufactured in accordance with this standard shall conform to two grades as follows:

1. **Grade N.** Units having a weight classification of 85 pcf or greater, for general use such as in exterior walls below and above grade that may or may not be exposed to moisture penetration or the weather and for interior walls and backup.

2. **Grade S.** Units having a weight classification of less than 85 pcf, for uses limited to above grade installation in exterior walls with weather-protective coatings and in walls not exposed to the weather.

(b) **Types.** Two types of concrete masonry units in each of two grades are covered as follows:

1. **Type 1, moisture-controlled units.** Units designated as Type I (Grades N–I and S–I) shall conform to all requirements of this standard including the moisture content requirements of Table No. 24-4-A.

2. **Type II, nonmoisture-controlled units.** Units designated as Type II (Grades N–II and S–II) shall conform to all requirements of this standard except the moisture-content requirements of Table No. 24-4-A.

Materials

Sec. 24.403. (a) **Cementitious Materials.** Materials shall conform to the following applicable standards:

1. Portland Cement—U.B.C. Standard No. 26-1. modified as follows:

> Limitation on insoluble residue—1.5 percent maximum
> Limitation on air content of mortar,
>> Volume Percent—22 percent maximum.
> Limitation on loss on ignition—7 percent maximum.
> Limestone with a minimum 85 percent calcium carbonate ($CaCO_3$) content may be added to the cement, provided the requirements of U.B.C. Standard No. 26-I as modified above are met.

2. Blended Cements—U.B.C. Standard No. 26-I.

3. Hydrated Lime, Type S—U.B.C. Standard No. 24-18.

4. Pozzolans—U.B.C. Standard No. 26-9.

(b) **Aggregates.** Aggregates shall conform to the following standards, except that grading requirements shall not necessarily apply:

1. Normal weight aggregates—U.B.C. Standard No. 26-2.

2. Lightweight aggregates—U.B.C. Standard No. 26-3.

(c) **Other Constituents.** Air-entraining agents, coloring pigments, integral water repellents, finely ground silica, etc., shall be previously established as suitable for use in concrete and either shall conform to U.B.C. Standard No. 26-9 where applicable, or shall be shown by test or experience to be not detrimental to the durability of the concrete.

Physical Requirements

Sec. 24.404. At the time of delivery to the work site the units shall conform to the physical requirements prescribed in Table No. 24-4-B. The moisture content of Type I concrete masonry units at time of delivery shall conform to the requirements prescribed in Table No. 24-4-A.

Minimum Face-shell and Web Thicknesses

Sec. 24.405. Face-shell (FST) and web (WT) thicknesses shall conform to the requirements listed in Table No. 24-4-C.

NOTE 3: Special unit designs involving corrosion-resistant metal ties between face shells may be approved, provided tests show such ties are structurally equivalent to the minimum specified concrete webs in stiffness, strength and anchorage to the face shells.

Permissible Variations in Dimensions

Sec. 24.406. (a) **Precision Units.** For precision units, no overall dimension (width, height and length) shall differ by more than $\frac{1}{8}$ inch from the specified standard dimensions.

(b) **Particular Feature Units.** For particular feature units, dimensions shall be in accordance with the following:

1. For molded face units, no overall dimension (width, height, and length) shall differ by more than $\frac{1}{8}$ inch from the specified standard dimension. Dimensions of molded features (ribs, scores, hex-shapes, patterns, etc.) shall be within 1/16 inch of the specified standard dimensions and shall be within 1/16 inch of the specified placement of the unit.

2. For split faced units, all non-split overall dimensions (width, height, and length) shall differ by no more than $\frac{1}{8}$ inch from the specified standard dimensions. On faces that are split, overall dimensions will vary. Local suppliers should be consulted to determine dimensional tolerances achievable.

3. For slumped units, no overall height dimension shall differ by more than $\frac{1}{8}$ inch from the specified standard dimension. On faces that are slumped, overall dimensions will vary. Local suppliers should be consulted to determine dimension tolerances achievable.

NOTE 4: Standard dimensions of units are the manufacturer's designated dimensions. Nominal dimensions of modular size units, except slumped units, are equal to the standard dimensions plus $\frac{3}{8}$ inch, the thickness of one standard mortar joint. Slumped units are equal to the standard dimensions plus $\frac{1}{2}$ inch, the thickness of one standard mortar joint. Nominal dimensions of nonmodular size units usually exceed the standard dimensions by $\frac{1}{8}$ to $\frac{1}{4}$ inch.

Visual Inspection

Sec. 24.407. (a) All units shall be sound and free of cracks or other defects that would interfere with the proper placing of the unit or impair the strength or permanence of the construction. Units may have minor cracks incidental to the usual method of manufacture or minor chipping resulting from customary methods of handling in shipment and delivery.

(b) Units that are intended to serve as a base for plaster or stucco shall have a sufficiently rough surface to afford a good bond.

(c) Where units are to be used in exposed wall construction, the face or faces that are to be exposed shall be free of chips, cracks or other imperfections when viewed from 20 feet, except that not more than 5 percent of a shipment may have slight cracks or small chips not larger than 1 inch.

Sampling and Testing

Sec. 24.408. The purchaser or his authorized representative shall be accorded proper facilities to inspect and sample the units at the place of manufacture from the lots ready for delivery. At least 10 days should be allowed for completion of the tests.

Sample and test units in accordance with U.B.C. Standard No. 24-7, Sampling and Testing Concrete Masonry Units.

When Type I moisture-controlled units are specified, moisture-content requirements (Table No. 24-4-A) shall be based upon U.B.C. Standard No. 24-26, Test Method for Drying Shrinkage of Concrete Block, not more than 12 months prior to delivery of units.

Rejection

Sec. 24.409. If the shipment fails to conform to the specified requirements, new specimens shall be selected from the retained lot and tested. If the second set of specimens fails to conform to the test requirements, the entire lot shall be rejected.

TABLE NO. 24-4-A—MOISTURE CONTENT REQUIREMENTS FOR TYPE I UNITS

LINEAR SHRINKAGE, PERCENT	MOISTURE CONTENT, MAX. PERCENT OF TOTAL ABSORPTION (Average of 3 Units)		
	Humidity Conditions at Jobsite or Point of Use		
	Humid[1]	Intermediate[2]	Arid[3]
0.03 or less	45	40	35
From 0.03 to 0.045	40	35	30
0.045 to 0.065, max.	35	30	25

[1]Average annual relative humidity above 75 percent.
[2]Average annual relative humidity 50 to 75 percent.
[3]Average annual relative humidity less than 50 percent.

TABLE NO. 24-4-B—STRENGTH AND ABSORPTION REQUIREMENTS

COMPRESSIVE STRENGTH, MIN., psi			WATER ABSORPTION, MAX. (Avg. of 3 units) WITH OVEN-DRY WEIGHT OF CONCRETE Lb./Cu. Ft.			
Average Net Area			Weight Classification			
			Lightweight		Medium Weight	Normal Weight
Grade	Avg. of 3 Units	Individual Unit	Less than 85	Less than 105	Less than 125 to 105	125 or more
N-I	1900	1500	—	18	15	13
N-II	1900	1500	—	18	15	13
S-I[1]	1300	1100	20	—	—	—
S-II[1]	1300	1100	20	—	—	—

[1]Limited to use above grade in exterior walls with weather-protective coatings and in walls not exposed to the weather.

NOTE: To prevent water penetration, protective coating should be applied on the exterior face of basement walls and when required on the face of exterior walls above grade.

TABLE NO. 24-4-C—MINIMUM THICKNESS OF FACE-SHELLS AND WEBS

NOMINAL WIDTH (W) OF UNITS, In.	FACE-SHELL THICKNESS (FST) MIN., In.[1] [4]	WEB THICKNESS (WT.)	
		Webs[1] Min., In.	Equivalent Web Thickness, Min. In./Lin.Ft.[2]
3 and 4	¾	¾	1⅝
6	1	1	2¼
8	1¼	1	2¼
10	1⅜	1⅛	2½
	1¼[3] [4]		
12	1½	1⅛	2½
	1¼[3] [4]		

[1]Average of measurements on three units taken at the thinnest point, when measured as described in U.B.C. Standard No. 24-7.

[2]Sum of the measured thickness of all webs in the unit, multiplied by 12 and divided by the length of the unit. In the case of open-ended units where the open-ended portion is solid grouted, the length of that open-ended portion shall be deducted from the overall length of the unit.

[3]This face-shell thickness (FST) is applicable where allowable design load is reduced in proportion to the reduction in thicknesses shown, except that allowable design load on solid-grouted units shall not be reduced.

[4]For split-faced units, a maximum of 10 percent of a shipment may have face-shell thicknesses less than those shown, but in no case less than ¾ inch.

UNIFORM BUILDING CODE STANDARD NO. 24-6
NONLOAD-BEARING CONCRETE MASONRY UNITS
Based on Standard Specification C 129-75 (1980) of the American Society for Testing and Materials
See Section 2402 (b) 5, Uniform Building Code

Scope

Sec. 24.601. This standard covers hollow and solid nonload-bearing concrete masonry units made from portland cement, water, and mineral aggregates with or without the inclusion of other materials. Such units are intended for use in nonload-bearing partitions but under certain conditions may be suitable for use in nonload-bearing exterior walls above grade, where effectively protected from the weather.

NOTE 1: Concrete masonry units covered by this standard are made from lightweight or normal weight aggregates, or both.

NOTE 2: When particular features are desired, such as weight classification, surface texture for appearance or bond, finish, color, fire resistance, insulation, acoustical properties, or other special features, such properties should be specified separately by the purchaser. However, local sellers should be consulted as to the availability of units having the desired features.

Classification

Sec. 24.602 (a) Weight Classifications. Nonload-bearing concrete masonry units manufactured in accordance with this standard shall conform to one of three weight classifications and two types as follows:

WEIGHT CLASSIFICATION	OVEN-DRY WEIGHT OF CONCRETE lb./ft.^{3}
Lightweight	105 max.
Medium weight	105–125
Normal weight	125 min.

(b) **Types.** Nonload-bearing concrete masonry units shall be of two types as follows:

1. **Type I, moisture-controlled units.** Type I units shall conform to all requirements of this standard, including the requirements of Table No. 24-6-A.

2. **Type II, nonmoisture-controlled units.** Type II units shall conform to all requirements of this standard, except the requirements listed in Table No. 24-6-A.

Materials

Sec. 24.603. (a) **Cementitious Materials.** Cementitious materials shall conform to the following applicable U.B.C. standards:

1. Portland Cement—U.B.C. Standard No. 26-1 modified as follows:

    Limitation on insoluble residue—1.5 percent

    Limitation on air content of mortar,

        Volume Percent—22 percent maximum.

    Limitation on loss on ignition—7 percent maximum.

    Limestone with a minimum 85 percent calcium carbonate ($CaCO_3$) content may be added to the cement, provided the requirements of U.B.C. Standard No. 26-1 as modified above are met.

2. Blended Cements—U.B.C. Standard No. 26-1.

3. Hydrated Lime, Type S—U.B.C. Standard No. 24-18.

4. Pozzolans—U.B.C. Standard No. 26-9.

(b) **Aggregates.** Aggregates shall conform to the following U.B.C. standards, except that grading requirements shall not necessarily apply:

1. Normal weight—U.B.C. Standard No. 26-2.

2. Lightweight—U.B.C. Standard No. 26-3.

(c) **Other Constituents.** Air-entraining agents, coloring pigments, integral water repellents, finely ground silica, etc., shall be previously established as suitable for use in concrete and either shall conform to U.B.C. Standard No. 26-9 where applicable or shall be shown by test or experience not to be detrimental to the durability of the concrete.

Physical Requirements

Sec. 24.604. At the time of delivery to the work site the units shall conform to the strength requirements prescibed in Table No. 24-6-B.

The moisture content of Type I concrete masonry units at the time of delivery shall conform to the requirements prescribed in Table No. 24-6-A.

Dimensions and Permissible Variations

Sec. 24.605. Minimum face-shell thickness shall be not less than ½ inch.

No overall dimension (width, height or length) shall not differ by more than ⅛ inch from the specified standard dimensions.

NOTE 3: Standard dimensions of units are the manufacturer's designated dimensions. Nominal dimensions of modular size units are equal to the standard dimensions plus ⅜ inch, the thickness of one standard mortar joint. Nominal dimensions of nonmodular size units usually exceed the standard dimensions by ⅛ to ¼ inch.

Variations in thickness of architectural units such as split-faced or slumped units will usually exceed the specified tolerances.

Visual Inspection

Sec. 24.606. (a) General. All units shall be sound and free of cracks or other defects that would interfere with the proper placing of the units or impair the strength or permanence of the construction. Units may have minor cracks incidental to the usual method of manufacture, or minor chipping resulting from customary methods of handling in shipment and delivery.

(b) Exposed Units. Where units are to be used in exposed wall construction, the face or faces that are to be exposed shall be free of chips, cracks or other imperfections, when viewed from 20 feet, except that not more than 5 percent of a shipment may have slight cracks or small chips not larger than 1 inch.

(c) Identification. Nonloadbearing concrete masonry units shall be clearly marked in a manner to preclude their use as load-bearing units.

Methods of Sampling and Testing

Sec. 24.607. (a) The purchaser or his authorized representative shall be accorded proper facilities to inspect and sample the units at the place of manufacture from the lots ready for delivery. At least 10 days should be allowed for the completion of the tests.

Sample and test units in accordance with U.B.C. Standard No. 24-7.

The moisture content requirements of Type I units shall be determined in accordance with U.B.C. Standard No. 24-27.

Rejection

Sec. 24.608. If the shipment fails to conform to the specified requirements, the manufacturer may sort it and new specimens shall be selected by the purchaser from the retained lot and tested. If the second set of specimens fails to conform to the test requirements, the entire lot shall be rejected.

TABLE NO. 24-6-A—MOISTURE CONTENT REQUIREMENTS FOR TYPE I UNITS

	MOISTURE CONTENT, MAX. PERCENT OF TOTAL ABSORPTION (Average of 3 Units)		
	Humidity[1] Conditions at Jobsite or Point of Use		
LINEAR SHRINKAGE, PERCENT	Humid	Intermediate	Arid
0.03 or less	45	40	35
From 0.03 to 0.045	40	35	30
0.045 to 0.065, max.	35	30	25

[1]Arid—Average annual relative humidity less than 50 percent.
Intermediate—Average annual relative humidity 50–75 percent.
Humid—Average annual relative humidity above 75 percent.

TABLE NO. 24-6-B—STRENGTH REQUIREMENTS

	COMPRESSIVE STRENGTH (Average Net Area) Min., psi
Average of 3 units	600
Individual units	500

UNIFORM BUILDING CODE STANDARD NO. 24-7

SAMPLING AND TESTING CONCRETE MASONRY UNITS

Based on Standard Methods C140-75 (1980) of the American Society for Testing and Materials

See Section 2402 (c) 2, Uniform Building Code

Scope

Sec. 24.701. These methods cover the sampling and testing of concrete masonry units for compressive strength, absorption, weight, moisture content and dimensions.

Sampling

Sec. 24.702. (a) **Test Specimens.** For purposes of test, full-size concrete masonry units shall be representative of the whole lot of units from which they are selected. Units for moisture content tests shall be protected from rain and other moisture until tested.

Three full-size units shall be tested within 72 hours after delivery to the laboratory, during which time they shall be stored continuously in normal room air.

Units of unusual size, shape or strength may be sawed into segments, some or all of which shall be tested individually in the same manner as prescribed for full-size units. The strength of the full-size units shall be considered as that which is calculated from the average measured strength of the segments.

(b) **Number of Specimens.** For the strength, absorption and moisture content determinations, six units shall be selected from each lot of 10,000 units or fraction thereof and 12 units from each lot of more than 10,000 and less than 100,000 units. For lots of more than 100,000 units, six units shall be selected from each 50,000 units or fraction thereof contained in the lot.

The number of specimens prescibed may be reduced by one half where strength tests only are required.

Compressive Test Apparatus

Sec. 24.703. The test machines shall have an accuracy of plus or minus 1.0 percent over the load range. The upper bearing shall be spherically seated, hardened metal block firmly attached at the center of the upper head of the machine. The center of the sphere shall lie at the center of the surface held in its spherical seat but shall be free to turn in any direction, and its perimeter shall have at least ¼-inch clearance from the head to allow for specimens whose bearing surfaces are not exactly parallel. The diameter of the bearing surface shall be at least 6 inches. A hardened metal bearing block may be used beneath the specimen to minimize wear of the lower platen of the machine. The bearing block surfaces intended for contact with the specimen shall have a hardness not less than 60 HRC (620 HB). These surfaces shall not depart from plane surfaces by more than 0.001 inch in any 6-inch dimension. When the bearing area of the spherical bearing block is not sufficient to cover the area of the specimen, a steel plate with surfaces machined to true planes within a plus or minus 0.001 inch in any 7-inch dimension, and with thickness equal to at least the distance from the edge of the spherical bearings to the most distant corner of the specimen, shall be placed between the spherical bearing block and the capped specimen.

Compressive Strength

Sec. 24.704. (a) **Capping Test Specimens.** Cap-bearing surfaces of units by one of the following methods:

1. **Sulfur and granular materials.** Spread evenly on a nonabsorbent surface that has been lightly coated with oil proprietary or laboratory prepared mixtures of 40 to 60 weight percent sulfur, the remainder being ground fire clay or other suitable inert material passing a No. 100 sieve, with or without a plasticizer. Heat the sulfur mixture in a thermostatically controlled heating pot to a temperature sufficient to maintain fluidity for a reasonable period of time after contact with the capping surface. Take care to prevent overheating and stir the liquid in the pot just before use. The capping surface shall be plane within 0.003 inch in 16 inches and shall be sufficiently rigid and so supported as not to be measurably deflected during the capping operation. Place four 1-inch square steel bars on the surface plate to form a rectangular mold approximately ½ inch greater in either inside dimension than the masonry unit. Fill the mold to a depth of ¼ inch with molten sulfur material. Bring the surface of the unit to be capped quickly into contact with the liquid and insert the specimen, holding it so that its axis is at right angles to the surface of the capping liquid. Allow the unit to remain undisturbed until

solidification is complete. Allow the cap to cool for minimum of two hours before testing the specimens. Patching of caps shall not be permitted. Remove imperfect caps and replace with new ones.

2. **Gypsum plaster capping.** Spread evenly on a nonabsorbent surface that has been lightly coated with oil, a neat paste of special high-strength plaster and water. Such gypsum plaster, when gauged with water at the capping consistency, shall have a compressive strength at a two-hour age of not less than 3500 psi when tested as 2-inch cubes. The casting surface plate shall conform to the requirements described in Subsection 1 above. Bring the surface of the unit to be capped into contact with the capping paste; firmly press down the specimen with a single motion, holding it so that its axis is at right angles to the capping surface. The average thickness of the cap shall be not more than ⅛ inch. Patching of caps shall not be permitted. Remove imperfect caps and replace with new ones. Age the caps for at least two hours before testing the specimens.

(b) **Procedure. 1. Position of specimens.** Test specimens with the centroid of their bearing surfaces aligned vertically with the center of thrust of the spherically seated steel bearing block of the testing machine. Except for special units intended for use with their cores in a horizontal direction, test all hollow concrete masonry units with their cores in a vertical direction. Test masonry units that are 100 percent solid and special hollow units intended for use with their hollow cores in a horizonal direction in the same direction as in service.

2. **Speed of testing.** Apply the load up to one half of the expected maximum load at any convenient rate, after which adjust the controls of the machine as required to give a uniform rate of travel of the moving head such that the remaining load is applied in not less than one nor more than two minutes.

(c) **Calculations and Report.** Take the compressive strength of a concrete masonry unit as the maximum load in pounds-force divided by the gross cross-sectional area of the unit in square inches. The gross area of a unit is the total area of a section perpendicular to the direction of the load, including areas within cells and within re-entrant spaces, unless these spaces are to be occupied in the masonry by portions of adjacent masonry.

Where a minimum compressive strength on the average net area as well as on the gross area is specified, calculate the maximum load in pounds-force divided by the average net area and include in the report.

Calculate the average percentage of net area of the unit as follows:

$$\text{Average net area, } \% = (A/B) \times 100$$
$$\text{Net volume } (A), \text{ ft}^3 = C/D$$
$$\text{Gross volume } (B), \text{ ft}^3 = (W \times H \times L)/1728$$
$$\text{Weight/ft}^3 (D) = C/(E\text{-}F) \times 62.4$$

WHERE:

A = net volume of unit, ft.^{3}
B = gross volume of unit, ft.^{3}
C = dry weight of unit, lb.
D = density, lb./ft.^{3}
W = width of unit, in.
H = height of unit, in.
L = length of unit, in.
E = wet weight of unit, lb.
F = suspended immersed weight of unit, lb.

Report the results to the nearest 10 psi separately for each unit and as the average for the three units.

Absorption

Sec. 24.705. (a) Procedure. 1. Saturation. Immerse the test specimens in water at room temperature at 60° to 80°F. for 24 hours. Weight the specimens while suspended by a metal wire and completely submerged in water. Remove from the water and allow to drain for one minute by placing them on a ⅜-inch or coarser wire mesh, removing visible surface water with a damp cloth, and immediately weigh.

2. Drying. Subsequent to saturation, dry all specimens in a ventilated oven at 212° to 239°F. for not less than 24 hours and until two successive weighings at intervals of two hours show an increment of loss not greater than 0.2 percent of the last previously determined weight of the specimen.

(b) **Calculations and Report. 1. Absorption.** Calculate the absorption as follows:

$$\text{Absorption, lb./ft.}^3 = [(A - B)(A - C)] \times 62.4$$
$$\text{Absorption, percent} = [(A - B)/B] \times 100$$

WHERE:

A = wet weight of unit, lb.
B = dry weight of unit, lb.
C = suspended immersed weight of unit, lb.

2. **Moisture Content.** Calculate the as-sampled moisture content as follows: Moisture content, %

$$= [(A - B)/(C - B)] \times 100$$

WHERE:

A = sampled weight of unit, lb.
B = dry weight of unit, lb.
C = wet weight of unit, lb.

3. **Report.** Report all results separately for each unit and as the average for the three units.

Measurements and Report

Sec. 24.706. Read individual measurements of the dimensions of each unit to the nearest division of the scale or caliper and record the average.

Measure length (L) on the longitudinal center line of each face, width W across the top and bottom bearing surfaces at midlength, and height H on both faces at midlength. Measure face-shell (FST) and web (WT) thicknesses at the thinnest point of each such element ½ inch above the mortar-bed plane. Where opposite face shells differ in thickness by less than ⅛ inch, average their measurements. Disregard sash grooves, dummy joints and similar details in the measurements.

The report shall show the average length, width and height of each specimen and the minimum face-shell and web thickness and the equivalent web thickness as an average for the three specimens.

UNIFORM BUILDING CODE STANDARD NO. 24-15
JOINT REINFORCEMENT FOR MASONRY
Specification Standard of the International Conference of Building Officials
See Sections 2402 (b) 10 and 2404 (i), Uniform Building Code
Part I—Joint Reinforcement for Masonry

Scope

Sec. 24.1501. This standard covers joint reinforcement fabricated from cold-drawn steel wire for reinforcing masonry.

Description

Sec. 24.1502. Joint reinforcement consists of deformed longitudinal wires welded to cross wires (Figure No. 24-15-1) in sizes suitable for placing in mortar joints between masonry courses.

Configuration and Size of Longitudinal and Cross Wires

Sec. 24.1503. (a) **General.** The distance between longitudinal wires and the configuration of cross wires connecting the longitudinal wires shall conform to the design.

(b) **Longitudinal Wires.** The diameter of longitudinal wires shall not be less than 0.148 inch (No. 9 gauge) nor more than one-half the mortar joint thickness.

(c) **Cross Wires.** The diameter of cross wires shall not be less than (No. 9 gauge) 0.148 inch diameter nor more than the diameter of the longitudinal wires. Cross wires shall not project beyond the outside longitudinal wires by more than ⅛ inch.

(d) **Width.** The width of joint reinforcement shall be the out-to-out distance between outside longitudinal wires. Variation in the width shall not exceed ⅛ inch.

(e) **Length.** The length of pieces of joint reinforcement shall not vary more than ½ inch or 1.0 percent of the specified length, whichever is less.

Material Requirements

Sec. 24.1504. (a) **Tensile Properties.** Wire of the finished product shall meet the following requirements:

Tensile strength, minimum	75,000 psi
Yield strength, minimum	60,000 psi
Reduction of area, minimum	30 percent

For wire testing over 100,000 psi, the reduction of area shall not be less than 25 percent.

(b) **Bend Properties.** Wire shall not break or crack along the outside diameter of the bend when tested in accordance with section 24.1508.

(c) **Weld Shear Properties.** The least weld shear strength in pounds shall be not less than 25,000 multiplied by the specified area of the smaller wire in square inches.

Fabrication

Sec. 24.1505. Wire shall be fabricated and finished in a workmanlike manner, shall be free from injurious imperfections and shall conform to this standard.

The wires shall be assembled by automatic machines or by other suitable mechanical means which will assure accurate spacing and alignment of all members of the finished product.

Longitudinal and cross wires shall be securely connected at every intersection by a process of electric-resistance welding.

Longitudinal wires shall be deformed. One set of four deformations shall occur around the perimeter of the wire at a maximum spacing of 0.7 times the diameter of the wire but not less than eight sets per inch of length. The overall length of each deformation within the set shall be such that the summation of gaps between the ends of the deformations shall not exceed 25 percent of the perimeter of the wire. The height or depth of the deformations shall be 0.012 inch for 3/16 inch diameter or larger wire, 0.011 for No. 8 gauge wire (0.162-inch diameter) and 0.009 inch for No. 9 gauge wire (0.148-inch diameter).

Tension Tests

Sec. 24.1506. Tension tests shall be made on individual wires cut from the finished product across the welds.

Tension tests across a weld shall have the welded joint located approximately at the center of the wire being tested.

Tensile strength shall be the average of four test values determined by dividing the maximum test load by the specified cross-sectional area of the wire.

Reduction of area shall be determined by measuring the ruptured section of a specimen which has been tested.

Weld Shear Strength Tests

Sec. 24.1507. Test specimens shall be obtained from the finished product by cutting a section of wire which includes one weld.

Weld shear strength tests shall be conducted using a fixture of such design as to prevent rotation of the cross wire. The cross wire shall be placed in the anvil of the testing device which is secured in the tensile machine and the load then applied to the longitudinal wire.

Weld shear strength shall be the average test load in pounds of four tests.

Bend Tests

Sec. 24.1508. Test specimens shall be obtained from the finished product by cutting a section of wire without welds.

The test specimens shall be bent cold through 180 degrees around a pin, the diameter of which is equal to the diameter of the specimen.

The specimen shall not break nor shall there be visual cracks on the outside diameter of the bend.

Frequency of Tests

Sec. 24.1509. One set of tension tests, weld strength shear tests and bend tests shall be performed for each 2,000,000 lineal feet of joint reinforcement but not less than monthly.

Corrosion Protection

Sec. **24.1510.** When corrosion protection of joint reinforcement is provided, it shall be in accordance with one of the following:

1. **Brite Basic.** No coating.

2. **Mill Galvanized.** Zinc coated, by the hot-dipped method, with no minimum thickness of zinc coating. The coating may be applied before fabrication.

3. **Class I Mill Galvanized.** Zinc coated, by the hot-dipped method, with a minimum of 0.40 ounces of zinc per square foot of surface area. The coating may be applied before fabrication.

4. **Class III Mill Galvanized.** Zinc coated, by the hot-dipped method, with a minimum of 0.80 ounces of zinc per square foot of surface area. The coating may be applied before fabrication.

5. **Hot-dipped Galvanized.** Zinc coated, by the hot-dipped method, with a minimum of 1.50 ounces of zinc per square foot of surface area. The coating shall be applied after fabrication.

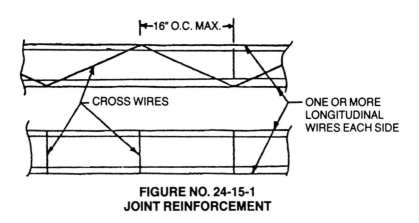

FIGURE NO. 24-15-1
JOINT REINFORCEMENT

COLD-DRAWN STEEL WIRE FOR
CONCRETE REINFORCEMENT

**Based on Standard Specification A82–85 of the American Society for
Testing and Materials**

**See Section 2402 (b) 10, Uniform Building Code
Part II**

Scope

Sec. 24.1511. This standard covers cold-drawn steel wire to be used
as such or in fabricated form, for the reinforcement as follows:

SIZE NUMBER	NOMINAL DIAMETER (inch)	NOMINAL AREA (Square inches)
W31	0.628	0.310
W30	0.618	0.300
W28	0.597	0.280
W26	0.575	0.260
W24	0.553	0.240
W22	0.529	0.220
W20	0.505	0.200
W18	0.479	0.180
W16	0.451	0.160
W14	0.422	0.140
W12	0.391	0.120
W10	0.357	0.100
W8	0.319	0.080
W7	0.299	0.070
W6	0.276	0.060
W5.5	0.265	0.055
W5	0.252	0.050
W4.5	0.239	0.045
W4	0.226	0.040
W3.5	0.211	0.035
W3	0.195	0.030
W2.5	0.178	0.025
W2	0.160	0.020
W1.5	0.138	0.015
W1.4	0.134	0.014
W1.2	0.124	0.012
W1	0.113	0.010
W0.5	0.080	0.005

Process

Sec. 24.1512. The steel shall be made by one or more of the following processes: open hearth, electric furnace or basic oxygen.

The wire shall be cold drawn from rods that have been hot rolled from billets.

Unless otherwise specified, the wire shall be "as cold drawn," except wire smaller than size number W1.2 for welded fabric, which shall be galvanized at finish size.

Tensile Properties

Sec. 24.1513. The material, except as specified in this section, shall conform to the following tensile property requirements based on nominal area of wire:

Tensile strength, minimum, psi 80,000
Yield strength, minimum, psi 70,000
Reduction of area, minimum, percent 30

For material testing over 100,000 pounds per square inch tensile strength, the reduction of area shall be not less than 25 percent.

For material to be used in the fabrication of welded fabric, the following tensile and yield strength properties based on nominal area of wire shall apply:

	SIZE W 1.2 AND LARGER	SMALLER THAN SIZE W 1.2
Tensile strength, minimum, psi	75,000	70,000
Yield strength, minimum, psi	65,000	56,000

The yield strength shall be determined at an extension of 0.005 inch per inch of gage length.

The material shall not exhibit a definite yield point as evidenced by a distinct drop of the beam or halt in the gage of the testing machine prior to reaching ultimate tensile load.

Bending Properties

Sec. 24.1514. The bend test specimen shall stand being bent cold through 180 degrees without cracking on the outside of the bent portion, as follows:

SIZE NUMBER OF WIRE	BEND TEST
W 7 and smaller	Bend around a pin, the diameter of which is equal to the diameter of the specimen.
Larger than W 7	Bend around a pin, the diameter of which is equal to twice the diameter of the specimen.

Test Specimens

Sec. 24.1515. Tension and bend test specimens shall be of the full section of the wire as drawn, or in the case of galvanized wire, as galvanized.

Number of Tests

Sec. 24.1516. One tension test and one bend test shall be made from each 10 tons or less of each size of wire.

If any test specimen shows imperfections or develops flaws, it may be discarded and another specimen substituted.

Permissible Variations in Wire Diameter

Sec. 24.1517. The permissible variation in the diameter of the wire shall conform to the following:

SIZE NUMBER	PERMISSIBLE VARIATION PLUS AND MINUS (Inch)
Smaller than W 5	0.003
W 5 to W 12, inclusive	0.004
Over W 12 to W 20, inclusive	0.006
Over W 20	0.008

The difference between the maximum and minimum diameter, as measured on any given cross section of the wire, shall be more than the tolerances shown above for the given wire size.

Finish

Sec. 24.1518. The wire shall be free from injurious imperfections and shall have a workmanlike finish with smooth surface.

Galvanized wire shall be completely covered in a workmanlike manner with a zinc coating.

UNIFORM BUILDING CODE STANDARD NO. 24-16
CEMENT, MASONRY

Based on Standard Specification C91-83a of the American Society for Testing and Materials

See Section 2402 (b) 2, Uniform Building Code

Scope

Sec. 24.1601. This standard covers masonry cement for use in masonry mortars.

Physical Requirements

Sec. 24.1602. Masonry cement shall conform to the requirements set forth in Table No. 24-16-A.

TABLE NO. 24-16-A—PHYSICAL REQUIREMENTS

MASONRY CEMENT TYPE	N	S	M
Fineness, residue on a No. 325 sieve, maximum, percent	24	24	24
Soundness:			
Autoclave expansion, maximum, percent	1.0	1.0	1.0
Time of setting, Gilmore method:			
Initial set, minimum, hour	2	1½	1½
Final set, maximum, hour	24	24	24
Compressive strength (average of 3 cubes): The compressive strength of mortar cubes, composed of 1 part cement and 3 parts blended sand (half Graded Ottawa sand, and half standard 20-30 Ottawa sand) by volume, prepared and tested in accordance with this specification shall be equal to or higher than the values specified for the ages indicated below:			
7 days, psi	500	1300	1800
28 days, psi	900	2100	2900
Air content of mortar, prepared and tested in accordance with requirements of these applicable specifications:			
Minimum percent by volume	12	12	12
Maximum percent by volume	22	20	20
Water retention, flow after suction, minimum, percent of original flow	70	70	70

UNIFORM BUILDING CODE STANDARD NO. 24–17
QUICKLIME FOR STRUCTURAL PURPOSES
Based on Standard Specification C 5–79 (1984)
of the American Society for Testing and Materials
See Section 2402 (b) 3, Uniform Building Code

Scope

Sec. 24.1701. This standard covers all classes of quicklime, such as crushed lime, granular lime, ground lime, lump lime, pebble lime and pulverized lime, used for structural purposes.

General Requirements

Sec. 24.1702. Quicklime shall be slaked and aged in accordance with the printed directions of the manufacturer. The resulting lime putty shall be stored until cool.

Chemical Composition

Sec. 24.1703. The quicklime shall conform to the following requirements as to chemical composition, calculated to the nonvolatile basis:

	CALCIUM LIME	MAGNESIUM LIME
Calcium oxide, minimum, percent ..	75	—
Magnesium oxide, minimum, percent	—	20
Calcium and magnesium oxides, minimum, percent............	95	95
Silica, alumina, and oxide of iron, maximum, percent	5	5
Carbon dioxide, maximum, percent:		
If sample is taken at the place of manufacture	3	3
If sample is taken at any other place	10	10

Residue

Sec. 24.1704. The quicklime shall contain not more than 15 percent by weight of residue.

UNIFORM BUILDING CODE STANDARD 24-18
HYDRATED LIME FOR MASONRY PURPOSES
Based on Standard Specification C 207-79 (Reapproved 1984) of the American Society for Testing and Materials
See Section 2402 (b) 3, Uniform Building Code

Scope

Sec. 24.1801. This standard covers four types of hydrated lime. Types N and S are suitable for use in mortar, in the scratch and brown coats of cement plaster, for stucco, and for addition to portland-cement concrete. Types NA and SA are air-entrained hydrated limes that are suitable for use in any of the above uses where the inherent properties of lime and air entrainment are desired. The four types of lime sold under this specifications shall be designated as follows:

Type N—Normal hydrated lime for masonry purposes.

Type S—Special hydrated lime for masonry purposes.

Type NA—Normal air-entraining hydrated lime for masonry purposes.

Type SA—Special air-entraining hydrated lime for masonry purposes.

Note: Type S, special hydrated lime, and Type SA, special air-entraining hydrated lime, are differentiated from Type N, normal hydrated lime, and Type NA, normal air-entraining hydrated lime, principally by their ability to develop high, early plasticity and higher water retentivity and by a limitation on their unhydrated oxide content.

Definitions

Sec. 24.1802. HYDRATED LIME. The hydrated lime covered by Types N or S in this standard shall contain no additives for the purpose of entraining air. The air content of cement-lime mortars made with Types N or S shall not exceed 7 percent. Types NA and SA shall contain an air-entraining additive as specified by Section 24.1805 of this standard. The air content of cement-lime mortars made with Types NA or SA shall have a minimum of 7 percent and a maximum of 14 percent.

Additions

Sec. 24.1803. Types NA and SA hydrated lime covered by this specification shall contain additives for the purpose of entraining air.

Manufacturer's Statement

Sec. 24.1804. Where required, the nature, amount and identity of the air-entraining agent used and of any processing addition that may have been used shall be provided, as well as test data showing compliance of such air-entraining addition.

Chemical Requirements

Sec. 24.1805 2. Hydrated lime for masonry purpose shall conform to the requirements as to chemical composition set forth in Table No. 24-18-A.

TABLE NO. 24-18-A—CHEMICAL REQUIREMENTS

	HYDRATE TYPES			
	N	NA PERCENT	S	SA
Calcium and magnesium oxides (nonvolatile basis), min.	95	95	95	95
Carbon dioxide (as-received basis), max.				
If sample is taken at place of manufacture	5	5	5	5
If sample is taken at any other place	7	7	7	7
Unhydrated oxides (as-received basis), max.			8	8

Residue, Popping and Pitting

Sec. 24.1806. The four types of hydrated lime for masonry purposes shall conform to one of the following requirements:

1. The residue retained on a No. 30 sieve shall not be more than 0.5 percent, or

2. If the residue retained on a No. 30 sieve is over 0.5 percent, the lime shall show no pops and pits when tested.

Plasticity

Sec. 24.1807. The putty made from Type S, special hydrate, or Type SA, special air-entraining hydrate, shall have a plasticity figure of not less than 200 when tested commencing within 30 minutes after mixing with water.

Water Retention

Sec. 24.1808. Hydrated lime mortar made with Type N (normal hydrated lime) or Type NA (normal air-entraining hydrated lime), after suction for 60 seconds, shall have a water-retention value of not less than 75 percent when tested in a standard mortar made from the dry hydrate or from putty made from the hydrate which has been soaked for a period of 16 to 24 hours.

Hydrated lime mortar made with Type S (special hydrated lime) or Type SA (special air-entraining hydrated lime), after suction for 60 seconds, shall have a water-retention value of not less than 85 percent when tested in a standard mortar made from the dry hydrate.

Special Marking

Sec. 24.1809. When Types NA or SA air-entraining hydrated lime are delivered in packages, the type under this specification and the words "air-entraining" shall be plainly indicated thereon or, in case of bulk shipments, so indicated on shipping notices.

UNIFORM BUILDING CODE STANDARD NO. 24-20
MORTAR FOR UNIT MASONRY AND REINFORCED MASONRY OTHER THAN GYPSUM

Based on Standard Specifications C161–44T and C270–59T of the American Society for Testing and Materials

See Section 2402 (b) 8, Uniform Building Code

Scope

Sec. 24.2001. These specifications cover the required properties of mortars determined by laboratory tests for use in the construction of reinforced brick masonry structures and unit masonry structures. Two alternative specifications are covered as follows:

1. **Property specifications.** Property specifications are those in which the acceptability of the mortar is based on the properties of the ingredients (materials) and the properties (water retention and compressive strength) of samples of the mortar mixed and tested in the laboratory.

2. **Proportion specifications.** Proportion specifications are those in which the acceptability of the mortar is based on the properties of the ingredients (materials) and a definite composition of the mortar consisting of fixed proportions of these ingredients.

Unless data are presented to show that the mortar meets the requirements of the physical property specifications, the proportion specifications shall govern. For field tests of grout and mortars see U.B.C. Standard No. 24–22.

Property Specifications

Materials

Sec. 24.2002. (a) **General.** Materials used as ingredients in the mortar shall conform to the requirements specified in the pertinent U.B.C. Standards.

(b) **Cemetitious Materials.** Cementitious materials shall conform to the following specifications:

1. **Portland cement.** Type I, IA, II, IIA, III or IIIA of U.B.C. Standard No. 26–1.

2. **Masonry cements.** U.B.C. Standard No. 24–16.

3. **Quicklime.** U.B.C. Standard No. 24–17.

4. **Hydrated lime.** U.B.C. Standard No. 24–18.

(c) **Aggregates.** U.B.C. Standard No. 24–21.

(d) **Water.** Water shall be clean and free of deleterious amounts of acids, alkalies or organic materials.

(e) **Admixtures or Mortar Colors.** Admixtures or mortar colors shall not be added to the mortar at the time of mixing unless provided for in the contract specifications and, after the material is so added, the mortar shall conform to the requirements of the property specifications.

Only pure mineral mortar colors shall be used.

(f) **Antifreeze Compounds.** No anitfreeze liquid, salts or other substances shall be used in the mortar to lower the freezing point.

(g) **Storage of Materials.** Cementitious materials and aggregates shall be stored in such a manner as to prevent deterioration or intrusion of foreign material. Any material that has become unsuitable for good construction shall not be used.

Mixing Mortar

Sec. 24.2003. All cementitious materials and aggregates shall be mixed for a minimum period of three minutes, with the amount of water required to produce the desired workabillity, in a drum-type batch mixer.

NOTE: Hand mixing of the mortar may be permitted on small jobs, with the written approval of the hand mixing procedure by the building official.

Mortar

Sec. 24.2004. (a) **Mortar for Unit Masonry.** Mortar conforming to the proportion specifications shall consist of a mixture of cementitious material and aggregate conforming to the requirements of Section 24.2002, and the measurement and mixing requirements of Section 24.2003, and shall be proportioned within the limits given in Table No. 24–20–B for each mortar type specified.

(b) **Mortar for Reinforced Masonry.** In mortar used for reinforced masonry the following special requirements shall be met: Sufficient water has been added to bring the mixture to a plastic state. The volume of aggregate in mortar shall be at least two and one-fourth times but not more than three times the volume of cemetitious materials.

(c) **Aggregate Ratio.** The volume of damp, loose aggregate in mortar used in brick masonry shall be not less than two and one-fourth times nor more than three times the total separate volumes of cementitious materials used.

(d) **Water Retention.** Mortar shall conform to the water retention requirements of Table No. 24–20–A.

(e) **Air Content.** Mortar shall conform to the air content requirements of Table No. 24–20–A.

Compressive Strength

Sec. 24.2005. The average compressive strength of three 2-inch cubes of mortar (before thinning) shall be not less than the strength given in Table No. 24–20–A for the mortar type specified.

Proportion Specifications

Materials

Sec. 24.2006. (a) **General.** Materials used as ingredients in the mortar shall conform to the requirements of Section 24.2002 and to the requirements of this section.

(b) **Masonry Cement.** Masonry cement shall conform to the requirements of U.B.C. Standard No. 24–16.

(c) **Hydrated Lime.** Hydrated lime shall conform to either of the two following requirements:

1. The total free (unhydrated) calcium oxide (CaO) and magnesium oxide (MgO) shall be not more than 8 percent by weight (calculated on the as-received basis for hydrates).

2. When the hydrated lime is mixed with portland cement in the proportion set forth in Table No. 24–20–B, the mixture shall give an autoclave expansion of not more than .50 percent.

Hydrated lime intended for use when mixed dry with other mortar ingredients shall have a plasticity figure of not less than 200 when tested 15 minutes after adding water.

(d) **Lime Putty.** Lime putty made from either quicklime or hydrated lime shall be soaked for a period sufficient to produce a plasticity figure of not less than 200 and shall conform to either the requirements for limitation on total free oxides of calcium and magnesium or the autoclave test specified for hydrated lime in Subsection (b).

Mortar

Sec. 24.2007. Mortar shall consist of a mixture of cementitious materials and aggregate conforming to the requirements specified in Section 24.2004, mixed in one of the proportions shown in Table No. 24-20-B, to which sufficient water has been added to reduce the mixture to a plastic state.

TABLE NO. 24-20-A—PROPERTY SPECIFICATIONS FOR MORTAR[1]

MORTAR	TYPE	AVERAGE COMPRESSIVE STRENGTH OF 2-INCH CUBES AT 28 DAYS (Min., psi)	WATER RETENTION (Min., percent)	AIR CONTENT (Max., percent)[2]	AGGREGATE MEASURED IN A DAMP LOOSE CONDITION
Cement-lime	M	2500	75	12	Not less than 2¼ and not more than 3½ times the sum of the separate volumes of cementitious materials.
	S	1800	75	12	
	N	750	75	14[3]	
	O	350	75	14[3]	
Masonry cement	M	2500	75	18	
	S	1800	75	18	
	N	750	75	18	
	O	350	75	18	

[1] Laboratory-prepared mortar only.
[2] Determined in accordance with applicable standards.
[3] When structural reinforcement is incorporated in cement-lime mortar, the maximum air content shall be 12 percent.

TABLE NO. 24–20–B—PROPORTION SPECIFICATIONS
FOR MORTAR

MORTAR	TYPE	PORTLAND CEMENT OR BLENDED CEMENT[1]	MASONRY CEMENT[2]			HYDRATED LIME OR LIME PUTTY[1]	AGGREGATE MEASURED IN A DAMP, LOOSE CONDITION
			M	S	N		
Cement-lime	M	1	—	—	—	¼	Not less than 2¼ and not more than 3 times the sum of the separate volumes of cementitious materials.
	S	1	—	—	—	over ¼ to ½	
	N	1	—	—	—	over ½ to 1¼	
	O	1	—	—	—	over 1¼ to 2½	
Masonry cement	M	1	—	—	1	—	
	M	—	1	—	—	—	
	S	½	—	—	1	—	
	S	—	—	1	—	—	
	N	—	—	—	1	—	
	O	—	—	—	1	—	

[1] When plastic cement is used in lieu of portland cement, hydrated lime or putty may be added, but not in excess of one tenth of the volume of cement.

[2] Masonry cement conforming to the requirements of U.B.C. Standard No. 24–16.

UNIFORM BUILDING CODE STANDARD NO. 24-21
AGGREGATE FOR MASONRY MORTAR
Based on Standard Specification C 144-81 of the
American Society for Testing and Materials
See Section 2402 (b) 1, Uniform Building Code

Scope

Sec. 24.2101. This standard covers aggregate for use in masonry mortar.

Material

Sec. 24.2102. Aggregate for use in masonry mortar shall consist of natural sand or manufactured sand. Manufactured sand is the product obtained by crushing stone, gravel or air-cooled iron blast-furnace slag.

Grading

Sec. 24.2103. Aggregate for use in masonry mortar shall be graded within the limits shown in Table No. 24-21-A.

EXCEPTION: When an aggregate fails the gradation limits of this section, it may be used provided the mortar can be prepared to comply with the aggregate ratio, water retention and compressive strength requirements of Section 24.2004 (c) and (d) and Section 24.2005 of U.B.C. Standard No. 24-20.

TABLE NO. 24-21-A—AGGREGATE FOR MASONRY MORTAR

SIEVE SIZE	PERCENT PASSING	
	NATURAL SAND	MANUFACTURED SAND
No. 4 (4750-micron)	100	100
No. 8 (2360-micron)	95 to 100	95 to 100
No. 16 (1180-micron)	70 to 100	70 to 100
No. 30 (600-micron)	40 to 75	40 to 75
No. 50 (300-micron)	10 to 35	20 to 40
No. 100 (150-micron)	2 to 15	10 to 25
No. 200 (75-micron)	—	0 to 10

The aggregate shall not have more than 50 percent retained between any two consecutive sieves shown in the table above. There shall be not more than 25 percent retained between the No. 50 and the No. 100 sieve.

The fineness modulus shall not vary by more than .20 from the value assumed in selecting proportions for the mortar, unless suitable adjustments are made in proportions to compensate for the change in grading.

Deleterious Substances

Sec. 24.2104. The amount of deleterious substances in aggregate for masonry mortar, each determined on independent samples complying with the grading requirements of Section 24.2103, shall not exceed the following maximum permissible percentage by weight:

Clay lumps.................................... 1.0
Lightweight particles, other than blast-furnace slag,
floating on liquid having a specific gravity of 2.0... 0.5

Organic Impurities

Sec. 24.2105. The aggregate shall be free of injurious amounts of organic impurities. Except as herein provided, aggregates subjected to the test for organic impurities and producing a color darker than the standard shall not be used.

EXCEPTIONS: 1. Aggregate failing in the test may be used, provided that the discoloration is due principally to the presence of small quantities of coal, lignite or similar discrete particles.

2. Aggregate failing in the test may be used provided that when tested for mortar-making properties the mortar develops a compressive strength at seven and 28 days of not less than 95 percent of that developed by a similar mortar made from another portion of the same sample which has been washed in a 3 percent solution of sodium hydroxide followed by thorough rinsing in water. The treatment shall be sufficient so that the washed material produces a color lighter than the standard.

Soundness

Sec. 24.2106. Except as herein provided aggregate subjected to five cycles of the soundness test shall show a loss, weighted in accordance with the grading of a sample complying with the limitations set forth in this standard, not greater than 10 percent when sodium sulfate is used or 15 percent when magnesium sulfate is used.

Aggregate failing to meet the requirements of this section may be accepted provided that mortar of comparable properties made from similar aggregate from the same source has been exposed to weathering similar to that to be encountered for a period of more than five years without appreciable disintegration.

UNIFORM BUILDING CODE STANDARD NO. 24-22
FIELD TEST SPECIMENS FOR MORTAR
Test Standard of the International Conference of Building Officials
See Section 2405 (c) 3, Uniform Building Code

Field Compressive Test Specimens for Mortar

Sec. 24.2201. Spread mortar on the masonry units ½ inch to ⅝ inch thick and allow to stand for one minute, then remove mortar and place in a 2-inch by 4-inch cylinder in two layers, compressing the mortar into the cylinder using a flat-end stick or fingers. Lightly tap mold on opposite sides, level off and immediately cover molds and keep them damp until taken to the laboratory. After 48 hours' set, have the laboratory remove molds and place them in the fog room until tested in the damp condition.

Requirements

Sec. 24.2202. Each such mortar test specimen shall exhibit a minimum ultimate compressive strength of 1500 pounds per square inch.

UNIFORM BUILDING CODE STANDARD NO. 24–23
AGGREGATES FOR MASONRY GROUT
Based on Standard Specification C 404–76 (Reapproved 1981) of the American Society for Testing and Materials
See Section 2402 (b) 1, Uniform Building Code

Scope

Sec. 24.2301. This standard covers aggregate for use in grout for reinforced masonry.

General Characteristics

Sec. 24.2302. Aggregate shall consist of natural sand or manufactured sand, used alone or in combination with coarse aggregate as described in this standard. Manufactured sand is the product obtained by crushing stone, gravel, or air-cooled iron blast-furnace slag. Coarse aggregate shall consist of crushed stone, gravel, or air-cooled iron blast-furnace slag specially processed to assure suitable particle shape as well as gradation.

Grading

Sec. 24.2303. The grading shall conform to the requirements set forth in Table No. 24–23–A.

TABLE NO. 24–23–A—GRADING REQUIREMENTS

SIEVE SIZE	AMOUNTS FINER THAN EACH LABORATORY SIEVE (Square Openings), Percent by Weight				
	Fine Aggregate			Coarse Aggregate	
	Size No. 1	Size No. 2			
		Natural	Manufactured	Size No. 8	Size No. 89
½-inch	—	—	—	100	100
⅜-inch	100	—	—	85 to 100	90 to 100
No. 4 (4.76-mm)	95 to 100	100	100	10 to 30	20 to 55
No. 8 (2.38-mm)	80 to 100	95 to 100	95 to 100	0 to 10	5 to 30
No. 16 (1.19-mm)	50 to 85	60 to 100	60 to 100	0 to 5	0 to 10
No. 30 (595-micron)	25 to 60	35 to 70	35 to 70	—	0 to 5
No. 50 (297-micron)	10 to 30	15 to 35	20 to 40	—	—
No. 100 (149-micron)	2 to 10)	2 to 15	10 to 25	—	—
No. 200 (74-micron)	—	—	0 to 10	—	—

Deleterious Substances

Sec. 24.2304. The amounts of deleterious substances in either fine or coarse aggregate shall not exceed the following:

DELETERIOUS SUBSTANCES	PERMISSIBLE CONTENT (Maximum percent by Weight)
Friable particles	1.0
Lightweight particles, other than blast furnace slag, floating on liquid having a specific gravity of 2.0	0.5

Organic Impurities

Sec. 24.2305. The fine aggregate shall be free from injurious amounts of organic impurities. Except as herein provided, aggregates subjected to the test for organic impurities and producing a color darker than the standard shall not be used.

> **EXCEPTIONS:** 1. Fine aggregate failing in the test may be used, provided that the discoloration is due principally to the presence of small quantities of coal, lignite or similar discrete particles.
>
> 2. Fine aggregate failing in the test may be used, provided that when tested for mortar-making properties, the mortar develops a compressive strength at seven and 28 days of not less than 95 percent of that developed by a similar mortar made from another portion of the same sample which has been washed in a 3 percent solution of sodium hydroxide followed by thorough rinsing in water. The treatment shall be sufficient so that the washed material produces a color lighter than the standard.

Soundness

Sec. 24.2306. Except as herein provided, either fine or coarse aggregates subjected to five cycles of the soundness test shall show a loss, weighted in accordance with the grading of a sample complying with the limitations prescribed in Section 24.2303, not greater than 10 percent when sodium sulfate is used or 15 percent when magnesium sulfate is used.

Aggregate failing to meet the requirements of this section may be accepted, provided that grout of comparable properties made from similar aggregate from the same source has been exposed to weathering, similar to that to be encountered, for a period of more than five years without appreciable disintegration.

UNIFORM BUILDING CODE STANDARD NO. 24-26
TEST METHOD FOR COMPRESSIVE STRENGTH OF MASONRY PRISMS
Based on Standard Test Method E 447-80 of the American Society for Testing and Materials

Scope

Sec. 24.2601. This standard covers procedures for masonry prism construction, testing and procedures for determining the compressive strength of masonry.

Sec. 24.2602. Prisms shall be constructed on a flat, level base. Masonry units used in the prism shall be representative of the units used in the corresponding construction. Each prism shall be built in an opened moisture-tight bag which is large enough to enclose and seal the completed prism. The orientation of units, where top and bottom cross sections vary due to taper of the cells, or where the architectural surface of either side of the unit varies, shall be the same orientation as used in the corresponding construction. Prisms shall be a single wythe in thickness and laid up in stack bond (see Figure No. 24-26-1).

The length of masonry prisms may be reduced by saw cutting; however, prisms composed of regular shaped hollow units shall have at least one complete cell with one full width cross web on either end. Prisms composed of irregular-shaped units shall be cut to obtain as symmetrical a cross section as possible. The minimum length of saw-cut prisms shall be 4 inches.

Masonry prisms shall be laid in a full mortar bed (mortar bed both webs and face shells). Mortar shall be representative of that used in the corresponding construction. Mortar joint thickness, the tooling of joints and the method of positioning and aligning units shall be representative of the corresponding construction.

Prisms shall be a minimum of two units in height, but not less than 12 inches. Immediately following the construction of the prism, the moisture-tight bag shall be drawn around the prism and sealed.

Where the corresponding construction is to be solid grouted, prisms shall be solid grouted. Grout shall be representative of that used in the corresponding construction. Grout shall be placed not less than one day nor more than two days following the construction of the prism. Grout consolidation shall be representative of that used in the construction. Additional grout shall be placed in the prism after reconsolidation and settlement due to water loss, but prior to the grout setting. Excess grout shall be screeded off level with the top of prism. Where open-end units are used, additional masonry units shall be used as forms to confine the grout during placement. Masonry unit forms shall be sufficiently braced to prevent displacement during grouting. Immediately following the grouting operation, the moisture-tight bag shall be drawn around the prism and resealed.

Where the corresponding construction is to be partially grouted, two sets of prisms shall be constructed; one set shall be grouted solid and the other set shall not be grouted.

Where the corresponding construction is of multiwythe composite masonry, masonry prisms representative of each wythe shall be built and tested separately.

Prisms shall be left undisturbed for at least two days after construction.

Transporting Masonry Prisms

Sec. 24.2603. Prior to transporting each prism, strap or clamp the prism together to prevent damage during handling and transportation. Secure prism to prevent jarring, bouncing or falling over during transporting.

Curing

Sec. 24.2604. Prisms shall remain sealed in the moisture-tight bag until two days prior to testing. The moisture-tight bag shall then be removed and curing continued in laboratory air maintained at a temperature of 75°F. plus or minus 15°F. Prisms shall be tested at 28 days after constructing the prism or at test age designated.

Preparation for Testing

Sec. 24.2605. (a) Capping the Prism. Cap top and bottom of the prism prior to testing with sulfur-filled capping or with high-strength gypsum plaster capping (such as "Hydrostone" or "Hyprocal White"). Sulfur-filled capping material shall be 40 to 60 percent by weight sulfur, the remainder being ground fireclay or other suitable inert material passing a No. 100 sieve, with or without a plasticizer.

Spread the capping material over a level surface which is plane within 0.003 inch in 16 inches. Bring the surface to be capped into contact with the capping paste; firmly press down the specimen, holding it so that its axis is at right angles to the capping surfaces. The average thickness of the cap shall not exceed ⅛ inch. Allow caps to age at least two hours before testing.

(b) **Measurement of the Prism.** Measure the length and thickness of the prism to the nearest 0.01 inch by averaging three measurements taken at the center and quarter points of the height of the specimen. Measure the height of the prism, including caps, to the nearest 0.1 inch.

Test Procedure

Sec. 24.2606. (a) **Test Apparatus.** The test machine shall have an accuracy of plus or minus 1.0 percent over the load range. The upper bearing shall be spherically seated, hardened metal block firmly attached at the center of the upper head of the machine. The center of the sphere shall lie at the center of the surface held in its spherical seat, but shall be free to turn in any direction, and its perimeter shall have at least ¼-inch clearance from the head to allow for specimens whose bearing surfaces are not exactly parallel. The diameter of the bearing surface shall be at least 5 inches. A hardened metal bearing block may be used beneath the specimen to minimize wear of the lower platen of the machine. The bearing block surfaces intended for contact with the specimen shall have a hardness not less than 60 HRC (620HB). These surfaces shall not depart from plane surfaces by more than 0.001 inch in any 6-inch dimension. When the bearing area of the spherical bearing block is not sufficient to cover the area of the specimen, a steel plate with surfaces machined to true planes within plus or minus 0.001 inch in any 6 inches dimension, and with a thickness equal to at least the distance from the edge of the spherical bearings to the most distant corner, shall be placed between the spherical bearing block and the capped specimen.

(b) **Installing the Prism in the Test Machine.** Wipe clean the bearing faces of the upper and lower platens or bearing blocks and of the test specimen and place the test specimen on the lower platen or bearing block. Align both centroidal axes of the specimen with the center of thrust of the test machine. As the spherically seated block is brought to bear on the specimen, rotate its movable portion gently by hand so that uniform seating is obtained.

(c) **Loading.** Apply the load, up one half of the expected minimum load, at any convenient rate, after which adjust the controls of the machine so that the remaining load is applied at a uniform rate in not less than one nor more than two minutes.

(d) **Observations.** Describe the mode of failure as fully as possible or illustrate crack patterns, spalling, etc., on a sketch, or both. Note whether failure occurred on one side or one end of the prism prior to failure of the opposing side or end of the prism.

Calculations

Sec. 24.2607. Calculations of test results shall be as follows:

1. **Net Cross-sectional Area.** Determine the net cross-sectional area (square inches) of solid grouted prisms by multiplying the average measured width dimension (inches) by the average measured length dimension (inches). The net cross-sectional area of ungrouted prisms shall be taken as the net cross-sectional area of masonry units determined from representative sample of units in accordance with U.B.C. Standard No. 24–24 for clay masonry units.

2. **Masonry Prism Strength.** Determine the compressive strength of each prism (psi) by dividing the maximum compressive load sustained (pounds) by the net cross-sectional area of the prism (square inches).

3. **Compressive Strength of Masonry.** The compressive strength of masonry (psi) for each set of prisms shall be the lesser of the average strength of the prisms in the set, or 1.25 times the least prism strength multiplied by the prism height to thickness correction factor from Table No. 24–26–A. Where a set of grouted and nongrouted prisms are tested, the compressive strength of masonry shall be determined for grouted set and for the nongrouted set separately. Where a set of prisms is tested for each wythe of a multiwythe wall, the compressive strength of masonry shall be determined for each wythe separately.

TABLE NO. 24–26–A

Prisms h/t_p[1]	1.5	2.0	2.5	3.0	4.0	5.0
Correction factor	0.86	1.0	1.04	1.07	1.15	1.22

[1]h/t_p—Ratio of prism height to least actual lateral dimension of prism.

Masonry Prism Test Report

Sec. 24.2608. The test report shall include the following:

1. Name of testing laboratory and name of professional engineer responsible for the tests.

2. Designation of each prism tested and description of prism, including width, height and length dimensions, mortar-type, grout and masonry unit used in the construction.

3. Age of prism at time of test.

4. Maximum compressive load sustained by each prism, net cross-sectional area of each prism and net area compressive strength of each prism.

5. Test observations for each prism in accordance with Section 24.2606.

6. compressive strength of masonry for each set of prisms.

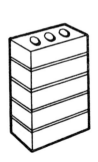 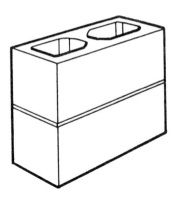

FIGURE NO. 24-26-1

UNIFORM BUILDING CODE STANDARD NO. 24-27
STANDARD TEST METHOD FOR
DRYING SHRINKAGE OF CONCRETE BLOCK
Based on Standard Test Method C 426-70 (Reapproved 1982) of The American Society for Testing and Materials

Scope

Sec. 24.2701. This method covers a routine standardized procedure for determining the drying shrinkage of concrete block, brick or other concrete products under specified accelerated drying conditions.

Definitions

Sec. 24.2702. DRYING SHRINKAGE in this method, is the change in linear dimension of the test specimen due to drying from a saturated condition to an equilibrium weight and length under specified accelerated drying conditions.

Apparatus

Sec. 24.2703. (a) **Measuring Instruments.** The instruments for measuring drying shrinkage shall be so designed as to permit or provide the following conditions:

NOTE 1: Strain gages may be obtained with various gage lengths. The 10-inch gage length is recommended for use with regular concrete block; however, particular sizes of products may require other lengths. The length of the shrinkage specimen shall not be less than required for a minimum gage length (distance between gage plugs) of 6 inches.

1. A means of positive contact with the specimen that will ensure reproducible measurements of length.

2. Means for precise measurement, consisting of a dial micrometer or other measuring device graduated to read in 0.0001-inch units, and accurate within 0.0001 inch in any 0.0010-inch range, and within 0.0002 inch in any 0.0100-inch range.

3. Sufficient range in the measuring device to allow for small variations in the gage lengths.

NOTE 2: If the shrinkage reference points are set carefully to position, a dial micrometer with a travel of 0.2 or 0.3 inch provides ample range in the instrument.

4. Means for checking the measuring device at regular intervals against a standard or reference.

NOTE 3: A standard reference bar shall be furnished by the manufacturer of the instrument. A standard bar of ordinary steel is satisfactory, but corrections must be made for variations in its length due to temperature changes. When a more nearly constant datum is desired, Invar is preferable because of its low coefficient of thermal expansion. The standard reference bar should be protected from air currents by placing it inside a wooden box, which should be closed except when the strain gage is being checked against it.

5. Convenient and rapid measurement of specimens.

(b) **Gage Plugs.** The gage plugs shall be made from metal that is resistant to corrosion. For use with strain gages, plugs shall be ⅜ to ½ inch in diameter and ½ ± ⅛ inch thickness.

1. As an alternative procedure, gage plugs for use with strain gages may be attached to the surface of the specimen with epoxy or polyester resin. These plugs shall be ⅜ to ½ inch in diameter and ⅜ ± ⅛ inch in thickness.

2. The method of application of the plugs to the surface of the specimens is described in Section 24.2705.

(c) **Drying Oven.** A reasonably airtight, insulated cabinet providing a minimum storage capacity of three whole block test specimens and capable of maintaining a constant temperature of 122 ± 2°F. One suggested oven construction is shown in Figure 24–27–1. The oven should provide the features described in 1 through 4.

1. A minimum of 1-inch clearance on all sides of each test specimen.

2. A constant, uniform temperature of 122 ± 2°F. throughout the cabinet attained by means of an electrical heat source (Note 4) and suitable thermal insulation.

NOTE 4: Direct heating of test specimens with the combustion products of gas or other carbonaceous fuels is not satisfactory due to the presence of carbon dioxide and water and their possible effect on the drying characteristics of portland cement products.

3. A means of drying specimens to a condition of euqilibrium with a relative humidity of 17 ± 2 percent (Note 5). Calcium chloride (CaCl₂), if used for this purpose, shall be in flake form.

NOTE 5: The air immediately above a saturated solution of calcium chloride (CaCl₂), at 122°F. is approximately 17 percent. Suitable dishes or trays shall be provided to give an exposed solution area of not less than 25 square inches for each cubic foot of oven volume. Dishes on trays shall contain sufficient solid calcium chloride so that the crystals will be exposed above the surface of the solution throughout the test. The calcium chloride solution shall be thoroughly stirred every 24 hours, and more often if necessary to prevent the formation of lumps and crusting over.

4. Moderate circulation of air within the oven, over and around all test specimens and the drying agent.

(d) **Cooling Chamber.** An airtight enclosure of sufficient capacity for cooling a minimum of three whole block test specimens.

(e) **Immersion Tank.** A suitable container for completely immersing three whole block test specimens in water maintained at 73.4 ± 2°F.

(f) **Balance or Scale.** The balance shall be sensitive to within 0.1 percent of the weight of the smallest specimen tested.

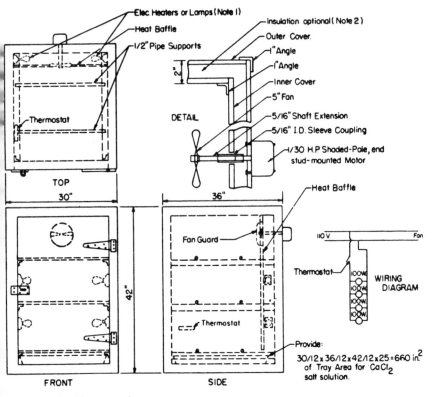

Figure labels (top-left / Top view): Elec. Heaters or Lamps (Note 1), Heat Baffle, 1/2" Pipe Supports, Thermostat, TOP, 30"

Detail labels: Insulation optional (Note 2), Outer Cover, 1" Angle, 1" Angle, Inner Cover, 5" Fan, 5/16" Shaft Extension, 5/16" I.D. Sleeve Coupling, 1/30 H.P. Shaded-Pole, end stud-mounted Motor, DETAIL, 2"

Front/Side view labels: Heat Baffle, 36", Fan Guard, 42", Thermostat, FRONT, SIDE

Wiring diagram labels: 110 V, Fan, Thermostat, 100W, 100W, 100W, 100W, WIRING DIAGRAM

Provide: $30/12 \times 36/12 \times 42/12 \times 25 = 660$ in.^{2} of Tray Area for $CaCl_2$ salt solution.

NOTE 1—Provide access to heaters.
NOTE 2—Insulating fill is recommended in cabinets having outer covers of sheet metal.
NOTE 3—The following materials are required:

Quantity	Description
1	5-in. fan assembly, as shown
1	1/30-hp (25-W) shaded-pole, fan-cooled, stud-mounted electric motor
75 ft.	1-in. angle, steel or aluminum
60 ft.^{2}	Outer cover, 1/2-in. plywood or equivalent, faced with sheet metal or other material to provide a positive vapor barrier
60 ft.^{2}	Inner cover, 3/8-in. asbestos board or equivalent
1	Heat baffle, 25 by 34-in. sheet metal
16 ft.	1/2-in. iron pipe
4	100-W porcelain light fixtures
1	500-W thermostat
1	24 x 30 x 1 1/2-in. tray, borosilicate glass or equivalent
1 pair	8-in. hinges and hasp

FIGURE NO. 24-27-1—Drying Oven Suitable for Determining Drying Shrinkage of Concrete Block

Test Specimens

Sec. 24.2704. The test specimens selected shall be whole units, free of visible cracks or other structural defects, which shall be representative of the lot from which they are selected (Note 6). Portions of face shells may be used in lieu of whole block providing they are dry cut lengthwise from both faces of hollow units at least 12 inches in length. Units previously subjected to tests involving temperatures exceeding 150°F. shall not be used in drying shrinkage tests.

NOTE 6: In tests of short units such as concrete brick by this method, use of a 10-inch Whittemore strain gage is reported to be feasible when two units are butted together and joined using an epoxy resin cement to form an extremely thin joint between the units. The abutting ends of the units should be ground to assure intimate contact and a thin joint: these precautions are necessary to ensure the thinnest joints practicable and thereby avoid abnormal shrinkage indications. Some laboratories have obtained satisfactory results using a 10-inch Whittemore strain gage on specimens joined with unfiled epoxy cements.

1. The number of specimens selected should consist of three whole units or six half-face shells.

2. The portions known as half-face shells should be at least 4 inches wide and should be of the same length as the face shell. They should be removed from the diagonally opposite ends of the two opposite face shells as shown diagrammatically in Figure No. 24-27-2. Half-face shell specimens must be dry cut from hollow units not less than 12 inches in length.

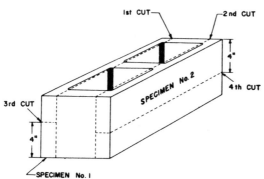

FIGURE NO. 24-27-2
View of Sawed Block Showing a Suggested Sequence of Cuts
and Location of Two Half-face Shell Specimens

Preparation of Apparatus

Sec. 24.2705. Place a pair of gage plugs at or near, and parallel to, the center line in each of two opposite faces of the whole unit specimen. Place a gage line in both the interior and exterior side of each half-face shell specimen. Drill plug holes with a drill that is slightly smaller in size than the plug diameter so as to provide a snug fit (Note 7). The depth of the holes shall be such that the exposed surface of the inserted gage plug is approximately 0.1 inch below the surface of the specimen. The cementing agent shall be portland cement paste, or an approved alternate cementing agent (Note 8). Prior to setting of the plug, plug holes shall be moistened when hydraulic cement is used, and shall be dry and dust free when plugs are set in epoxy or polyester resins or other nonhydraulic cementing materials. After the cementing material has been placed in the hole, insert the gage plug and prick punch the plug to proper gage length with the gage bar provided. Wipe off excessive cementing material and allow the remainder to cure. After the cementing material is sufficiently hard, drill receiving holes for strain gage points with a No. 56 to 60 twist drill.

NOTE 7: A 5/16-inch-diameter carbide-tipped masonry drill has been found satisfactory for gage plugs ⅜ inch in diameter.

NOTE 8: A number of cementing agents have been reported satisfactory for setting gage plugs. Materials such as dental cement and some of the polyester and epoxy resins make possible a considerable reduction in the preparation time of shrinkage test specimens. Tests to determine the effect of water immersion and subsequent drying on its adhesion should be made prior to the substitution of any cementing agent for portland cement.

When the gage plugs for use with strain gages are attached to the surface of concrete with epoxy or polyester resin, drill receiving holes for strain gage points prior to attachment of gage plugs. Attach gage plugs with epoxy or polyester resin (Note 8) using the strain gage punch bar or other convenient template to set gage holes the proper distance apart. The surface of metal gage plugs which will be in contact with resin should be roughened with emery cloth.

Procedure

Sec. 24.2706. The following procedure shall be used: 1. Immerse specimens for the drying shrinkage determination in water at 73.4 ± 2°F. for 48 hours.

2. Take the initial reading of specimen length, at saturation, with the unit positioned in the water tank so that its gage line is about at the level of the water surface to avoid error due to cooling by evaporation. Accompany length readings of test specimens by length readings of the standard reference bar.

3. Obtain the saturated surface-dry weight of the test specimen. A saturated surface-dry condition shall be obtained by draining the test specimen for 1 minute over a ⅜-inch (or larger) mesh and removing visible surface water by blotting with a damp cloth.

4. Store test specimens for drying in the oven described in Sec. 24.2703 (c). In exceptional circumstances it is permissible to surface dry the specimens in room air before storing in oven (Note 9). To ensure uniformity of drying, the individual specimens should be rotated to different positions in the drying oven each time readings are taken.

NOTE 9: Reports have indicated that moisture is exuded faster by some masonry units during the early part of the drying period than can be absorbed by the calcium chloride solution, causing condensation to form on the interior surfaces of the oven. Where this situation is encountered, it shall be permissible to allow units to lose excess water for periods up to 48 hours in room air before transfer to the drying oven, provided the effect of such deviation permitted under Item 4 above has been determined to be negligible.

5. At the end of five days of drying, including any period of preliminary drying in air up to 48 hours, remove shrinkage specimens from the drying oven and cool to 73.4 ± 2°F. (Note 10). Following cooling, obtain specimen length reading and weight and length reading of standard reference bar.

NOTE 10: Use a cooling chamber consisting of a steel, drum-type container equipped with a ring-sealed, rubber-gasketed type cover. The drum cover should be equipped with a thermometer, the bulb of which is in the proximity of the uppermost test specimen. The drum must be stored in a temperature-controlled room in order that its final equilibrium temperature will be 73.4°F. Length measurements made at temperatures other than 73.4°F. shall be corrected as shown in Section 24.2707(c).

(f) Return test specimens to the drying oven for a second period of drying. The duration of the second, and subsequent, drying periods shall be 48 hours. Following the second period of drying, repeat cooling, length readings, and weight determinations as specified in Section 24.2706(e).

7. Continue the 48-hour periods of drying in the specified oven, followed by length and weight determinations after cooling under the specified conditions (Note 10) until an equilibrium condition of the shrinkage specimens has been reached: equilibrium is considered to be the prevailing condition when the average length change of the test specimens is 0.002 percent, or less, over a span of six days of drying, and when the average weight loss in 48 hours of drying is 0.2 percent or less compared to the last previously determined weight.

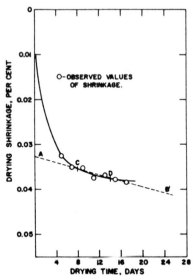

NOTE The interval *CD* is 6 days on the time scale and 0.002 percent on the shrinkage scale. Point *D* defines equilibrium shrinkage value.

FIGURE NO. 24–27–3
Graphical Method of Determining Equilibrium Shrinkage

NOTE 11: When uniform attainment of equilibrium length is not apparent in the tabular data, the value of equilibrium shrinkage may be obtained from shrinkage-time curves drawn through experimental points as illustrated in Figure No. 27-27-3. The dotted line AB having a slope corresponding to the limiting value of rate of shrinkage (0.002 percent in six days) is fitted to the experimental curve in such a manner that the points of intersection C and D span a time interval of six days. The corresponding shrikage interval between point C and D is 0.002 percent. The value of equilibrium shrinkage shall be taken as the shrinkage corresponding to point D expressed to the nearest 0.001 percent. Data for which the rate of shrinkage is obviously within the prescribed limit need not be plotted, but the principle of selecting point D should be followed. That is, the final percent shrinkage is the greater of two values agreeing within 0.002 percentage points over a period of six days.

Calculations

Sec. 24.2707. (a) **General.** Calculate the drying shrinkage as a percentage of the gage length as follows:

$$S = (L/G) \times 100$$

WHERE:

S = linear drying shrinkage, percent,

L = change in the linear dimension of the specimen due to drying from a saturated condition to the equilibrium weight and length as specfied in Sec. 24.2706, Item 7, and

G = test specimen gage length

(b) **Adjustment for Variation in Reference Bar Readings.** Adjust the reported change in linear dimension of the test specimen for variations in the reference bar readings that are due to causes other than temperature as follows:

$$L = (L_1 - R_1) - (L_2 - R_2)$$

WHERE:

L_1 = initial-length reading of the test specimen,

L_2 = final-length reading of the test specimen,

R_1 = initial-length reading of the reference bar, and

R_2 = final-length reading of the reference bar.

(c) **Method of Correcting Length Readings to Standard Temperature.** Correct length readings taken at temperatures other than 73.4°F. as follows:

$$L_{73.4} = L_x - (T_x - 73.4) \times GQ$$

WHERE:

L_x = length reading of specimen or reference bar taken at temperature T_x (Note 12),

G = gage length, and

Q = termal coefficient of expansion of material (Note 13).

NOTE 12: The variation of the temperature in the laboratory from the standard temperature of 73.4°F. at the time readings are made on the test specimens shall not exceed ± 5°F.

NOTE 13: The coefficient Q for mild steel is 0.0000065 in./in./°F. The coefficient Q for concrete, if unknown, may be assumed as 0.0000045 in./in./°F.

Report

Sec. 24.2708. The report shall include the following:

1. Identification of kind of product and number of specimens for each condition of test.
2. Source of specimens.
3. Kind of aggregate, type of portland cement, and method of producing the product.
4. Condition of curing and drying prior to test.
5. Age of specimens at start of shrinkage test.
6. Total length of drying period prior to each length measurement.
7. Weight of test specimens as received, saturated, and at the time of each length measurement, including equilibrium.
8. Total linear drying shrinkage, percent, from saturation to each length measurement, including the length measured at equilibrium.
9. Any other information that may be pertinent.

UNIFORM BUILDING CODE STANDARD NO. 24-28
METHOD OF SAMPLING AND TESTING GROUT
Based on Standard Method C 1019-84 of the American Society for Testing and Materials

Scope

Sec. 24.2801. This method covers procedures for both field and laboratory sampling and compression testing of grout used in masonry construction.

Apparatus

Sec. 24.2802. (a) **Maximum-Minimum Thermometer.**

(b) **Straightedge.** A steel straightedge not less than 6 inches long and not less than 1/16 inch in thickness.

(c) **Tamping Rod.** A nonabsorbent smooth rod, either round or square in cross section nominally 5/8 inch in dimension with ends rounded to hemispherical tips of the same diameter. The rod shall be a minimum length of 12 inches.

(d) **Wooden Blocks.** Wooden squares with side dimensions equal to one-half the desired grout specimen height, within a tolerance of 5 percent, and of sufficient quantity or thickness to yield the desired grout specimen height, as shown in Figures Nos. 24-28-A and 24-28-B.

Wooden blocks shall be soaked in limewater for 24 hours, sealed with varnish or wax, or covered with an impermeable material prior to use.

Sampling

Sec. 24.2803. (a) **Size of Sample.** Grout samples to be used for slump and compressive strength tests shall be a minimum of ½ ft.^{3}.

(b) **Field Sample.** Take grout samples as the grout is being placed into the wall. Field samples may be taken at any time except for the first and last 10 percent of the batch volume.

Test Specimen and Sample

Sec. 24.2804. (a) Each grout specimen shall be a square prism, nominally 3 inches or larger on the sides and twice as high as its width. Dimensional tolerances shall be within 5 percent of the nominal width selected.

(b) Three specimens shall constitute one sample.

Procedure

Sec. 24.2805. (a) Select a level location where the molds can remain undisturbed for 48 hours.

(b) **Mold Construction.** 1. The mold space should simulate the grout location in the wall. If the grout is placed between two different types of masonry units, both types should be used to construct the mold.

2. Form a square prism space, nominally 3 inches or larger on each side and twice as high as its width, by stacking masonry units of the same type and moisture condition as those being used in the construction. Place wooden blocks, cut to proper size and of the proper thickness or quantity, at the bottom of the space to achieve the necessary height of specimen. Tolerance on space and specimen dimensions shall be within 5 percent of the specimen width. See Figures Nos. 24–28–A and 24–28–B.

3. Line the masonry surfaces that will be in contact with the grout specimen with a permeable material, such as paper towel, to prevent bond to the masonry units.

(c) Measure and record the slump of the grout.

(d) Fill the mold with grout in two layers. Rod each layer 15 times with the tamping rod penetrating ½ inch into the lower layer. Distribute the strokes uniformly over the cross section of the mold.

(e) Level the top surface of the specimen with a straightedge and cover immediately with a damp absorbent material such as cloth or paper towel. Keep the top surface of the sample damp by wetting the absorbent material and do not disturb the specimen for 48 hours.

(f) Protect the sample from freezing and variations in temperature. Store an indicating maximum-minimum thermometer with the sample and record the maximum and minimum temperatures experienced prior to the time the specimens are placed in the moist room.

(g) Remove the masonry units after 48 hours. Transport field specimens to the laboratory, keeping the specimens damp and in a protective container.

(h) Store in a moist room conforming to nationally recognized standards.

(i) Cap the specimens in accordance with the applicable requirements of U.B.C. Standard No. 24–26.

(j) Measure and record the width of each face at midheight. Measure and record the height of each face at midwidth. Measure and record the amount out of plumb at midwidth of each face.

(k) Test the specimens in a damp condition in accordance with the applicable requirements of U.B.C. Standard No. 24-26.

Calculations

Sec. 24.2806. The report shall include the following:

(a) Mix design.

(b) Slump of the grout.

(c) Type and number of units used to form mold for specimens.

(d) Description of the specimens—dimensions, amount out of plumb—in percent.

(e) Curing history, including maximum and minimum temperatures and age of specimen, when transported to laboratory and when tested.

(f) Maximum load and compressive strength of the sample.

(g) Description of failure.

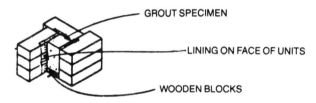

Note: Front masonry unit stack not shown to allow view of specimen.

FIGURE NO. 24-28-A
Grout Mold (Units 6 Inches or Less
In Height, 2¼-Inch-high Brick Shown)

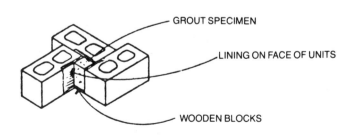

Note: Front masonry unit not shown to allow view of specimen.

UNIFORM BUILDING CODE STANDARD NO. 24–29
GROUT FOR MASONRY

Based on Standard Specification C 476-83 of the American Society for Testing and Materials
See Section 2402 (b) 9, Uniform Building Code

Scope

Sec. 24.2901. This standard covers grout for use in the construction of reinforced and nonreinforced masonry structures.

Materials

Sec. 24.2902. Materials used as ingredients in grout shall conform to the following:

1. **Cementitious Materials.** Cementitious materials shall conform to one of the following standards:

   A. Portland Cement—Types I, II and III of U.B.C. Standard No. 26-1.

   B. Blended Cement—Type IS, IS(MS) or IP of U.B.C. Standard No. 26-1.

   C. Quicklime—U.B.C. Standard No. 24-17.

   D. Hydrated Lime—Type S of U.B.C. Standard No. 24-18.

2. **Aggregates.** Aggregates shall conform to U.B.C. Standard No. 24-23.

3. **Water.** Water shall be clean and potable.

4. **Admixtures.** Additives and admixtures to grout shall not be used unless approved by the building official.

5. **Antifreeze Compounds.** No antifreeze liquids, chloride salts or other substances shall be used in grout.

6. **Storage of Materials.** Cementitious materials and aggregates shall be stored in such a manner as to prevent deterioration or intrusion of foreign material or moisture. Any material that has become unsuitable for good construction shall not be used.

Measurement of Materials

Sec. 24.2903. The method of measuring materials for the grout used in construction shall be such that the specified proportions of the grout materials can be controlled and accurately maintained.

Grout

Sec. 24.2904. Grout shall consist of cementitious material and aggregate that have been mixed thoroughly for a minimum of five minutes in a mechanical mixer with sufficient water to bring the mixture to the desired consistency. The grout proportions and any additives shall be based on laboratory or field experience considering the grout ingredients and the masonry units to be used, or the grout shall be proportioned within the limits given in Table No. 24-B of the Uniform Building Code, or the grout shall have a minimum compressive strength when tested in accordance with U.B.C. Standard No. 24-28 equal to its specified strength, but not less than 2000 psi.

References

1. *1988 Uniform Building Code* by International Conference of Building Officials

2. *1988 Uniform Building Code Standards*

3. *Reinforced Concrete Masonry Inspector's Manual* by Concrete Masonry Association of California and Nevada, 1983

4. *Concrete Masonry Design Manual* published by Concrete Masonry Association of California and Nevada, 1981

5. *Masonry Design Manual* by J.E. Amrhein and J.J. Kesler et al, 1979 published by Masonry Industry Advancement Committee

6. *Commentary to Chapter 24 of the Uniform Building Code 1985* published by The Masonry Society, 1986

7. *Modern Masonry Manual,* 1970

8. *Cement-Lime Mortars* by T.Ritchie and J.I. Davison—BRI Building Research, Mar/Apr 1964

9. *Masonry: Materials, Design, Construction* by R.C. Smith, T.L. Honkala and C.K. Andres, 1979, Prentice-Hall, Inc. Englewood Cliffs, N.J.

10. *Recommended Practices & Guide Specifications for Cold Weather Masonry Construction* by International Masonry Industry All Weather Council, Washington, D.C., 1970

11. *Advanced Masonry Skills* by Richard T. Kreh, Sr., 1978 Van Nostrand-Reinhold Co., New York City, N.Y.

12. *Masonry Design and Detailing* by Christine Beall, AIA, 1984 Prentice-Hall, Inc., Englewood Cliffs, N.J.

13. National Lime Association

14. *Why Masonry Walls Leak* by Walter C. Voss, Ph.D., published by National Lime Association, 1938.

15. Corps of Engineers Guide Specifications

16. California Youth Authority, Masonry, General Requirements Sections 04001, 04050, 04220, and 04221

17. *Sub-Committee Testing Guidelines for Construction Materials* by Nezih Gunal, P.E., Chairman, Report No. 2-Masonry and Masonry Materials, 1986, Structural Engineers Association of Southern California, Los Angeles, Calif.

18. *Cement Colors and Colored Concrete Products,* Williams, 1972 Technical Report, Pfizer Co., New York City, N.Y.

19. American Society for Testing and Materials:
    a. A82   Cold-Drawn Steel Wire for Concrete Reinforcement
    b. A615   Reinforcing Bars for Concrete
    c. C5   Quicklime for Structural Purposes
    d. C55   Concrete Building Brick
    e. C90   Hollow Load-Bearing Concrete Masonry Units
    f. C91   Masonry Cement
    g. C140   Sampling
    h. C144   Aggregate for Masonry Mortar
    i. C145   Solid Load-Bearing Concrete Masonry Units
    j. C150   Portland Cement
    k. C207   Hydrated Lime for Masonry Purposes
    l. C270   Mortar for Unit Masonry
    m. C404 Aggregate for Grout
    n. C476   Grout for Masonry
    o. C1019 Sampling and Testing Grout
    p. E447   Compressive Strength of Masonry Prisms

20. *Effect of Mortar Properties on the Flexural Bond Strength of Masonry* by G.C. Robinson; *Bonding of Brick to Mortar; Influence of the Type of Mortar and Air Content on Bond Strength,* Clemson University, Clemson, S. Carolina, 1986

21. *Reinforced Masonry Engineering Handbook,* by J.E. Amrhein, 1983, published by Masonry Institute of America, Los Angeles, Calif.

1988 MASONRY CODES and SPECIFICATIONS HANDBOOK

Edited by James E. Amrhein, S.E.

480 Pages of the Most Current Masonry Information

✔ 1988 Uniform Building Code
✔ 1988 Uniform Building Code Standards
✔ Quality Control Standards
✔ California Office of State Architect Requirements
✔ ASTM Standards for Masonry

Book Size: 5"x8"
ISBN 0-940116-11-1

$10.00

MARBLE AND STONE SLAB VENEER

SECOND EDITION

THE Source Book for Stone Slab Veneer Information

Covers all aspects of stone veneering an architect needs to know for the designing and detailing of stone slab veneer:

➢ **Design Requirements**
 ➢ **Tolerances**
  ➢ **Fabrication**
   ➢ **Panel Systems**
    ➢ **Detailed Illustrations**
   ➢ **Design Examples**
  ➢ **Code Requirements**
 ➢ **and much more...**

Book Size: 8½"x11" — 138 pages.
ISBN 0-940116-06-8

$15.00

RESIDENTIAL MASONRY
FIREPLACE AND CHIMNEY
HANDBOOK

by
James E. Amrhein, S.E.

Well illustrated information,
specially useful to architects,
designers, engineers, building
officials, construction inspectors,
masons and masonry contractors.

Includes:

- Fireplaces

- Clearances

- Sizes

- Dampers

- Chimneys

- Flues

- Reinforcing

- Specifications

- Care and operation

5½″ x 8″ publication
178 pages   **$10.00**

MASONRY
DESIGN
MANUAL

Includes:

- Specifications
- Typical designs
- Architectural details
- Brick
- Concrete Block
- Veneer
- General design
- Engineering
- Reinforcing steel details
- Retaining walls
- Noise control
      . . . .and more

Many details, design charts and tables necessary for the practicing architect. A must for the architectural designer!

Book Size: 8½ x 11"
360 pages

$23.80

ORDER BLANK
(other side)

ORDER FORM

Mail to:
Masonry Institute of America
2550 Beverly Blvd.
Los Angeles, CA 90057

NAME _____

STREET ADDRESS _____

_____

CITY _____ _____

STATE _____ **ZIP** _____

PHONE (_____ **) -** _____

(PLEASE, NO P.O. BOX NUMBER)

Publication	Unit Price	Qnty.	Total

TOTAL _____

A check or money order drawn on a U.S. bank payable in U.S. currency **MUST** accompany this order.

Orders from outside the United States will cost three (3) times the listed price.